Topics in Mining, Metallurgy and Materials Engineering

Series editor

Carlos P. Bergmann, Porto Alegre, Brazil

More information about this series at http://www.springer.com/series/11054

Hugo Marcelo Veit · Andréa Moura Bernardes
Editors

Electronic Waste

Recycling Techniques

Springer

Editors
Hugo Marcelo Veit
Departamento de Materiais, Escola de Engenharia
Universidade Federal do Rio Grande do Sul
Porto Alegre
Brazil

Andréa Moura Bernardes
Departamento de Materiais, Escola de Engenharia
Universidade Federal do Rio Grande do Sul
Porto Alegre
Brazil

ISBN 978-3-319-15713-9 ISBN 978-3-319-15714-6 (eBook)
Topics in Mining, Metallurgy and Materials Engineering
DOI 10.1007/978-3-319-15714-6

Library of Congress Control Number: 2015931538

Springer Cham Heidelberg New York Dordrecht London

Printed on acid-free paper

Springer International Publishing AG Switzerland is part of Springer Science+Business Media (www.springer.com)

To Prof. Dr.-Ing. Wolfgang Wuth, to the introduction to this fascinating topic and for the constant encouragement to the metals recycling subject.

Foreword

The current industrial society must solve two major problems: the lack of raw materials and the lack of energy. Most raw materials shortage is represented by metals. Unlike other materials, metal products can be replaced only to a limited extent or not at all. Global stocks of primary metal raw materials are getting depleted significantly and for some of them there are open deposits only for a couple of years. Therefore, in recent years the importance of metals recycling is increasing, not only in terms of sustainable development issues, but also because significant energy saving and environmental protection is achieved.

Worldwide, approximately 20–50 million tons of Waste Electrical and Electronic Equipments (WEEEs) are being produced annually. Such amount of waste is significant not only in terms of quantity, but also because it contains substances (heavy metals and plastics) that can have hazardous nature, especially when they are managed by incorrect or inappropriate way. On the other hand this waste represents a significant amount of valuable compounds, e.g.: 0.1 % Au, 0.2 % Ag, 20 % Cu, 4 % Sn. For comparison: the content of gold in a primary raw material is 2–5 g/t, which is 100 times less than in WEEEs. It follows that WEEEs should not be regarded as waste but as a valuable secondary raw material.

Most WEEEs include Printed Circuit Boards (PCBs). PCBs are normally separated from WEEEs, picked up by hand and sold to non-ferrous metal market because PCBs include a high grade of copper and other valuable metals. However, PCBs also contain other metals and have impurities such as silicon, aluminum, iron, which are slag materials in non-ferrous smelting process, and also other harmful elements such as zinc, lead, bromine, and antimony. The content of plastics in WEEEs is also not negligible.

The main difference between primary ore and waste, except quantitative representation of elements, is their presence of form. While metals in ores are mainly present as compounds, in WEEE they are usually present in metallic form or as alloys. This makes them suitable for specialized procedures of processing.

The present monograph complements the current state of published information about recycling of high-grade WEEE. It describes in compact form the basic operation steps of processing WEEE by physical, physicochemical, and chemical ways. The monograph also describes the basic steps of mechanical pretreatment and possibilities of hydrometallurgical, pyrometallurgical, and electrometallurgical procedures in recycling metals from selected categories of WEEE, mainly electronic waste, used batteries, lighting, and display technology. Many of the presented results are substantiated by the authors' own research.

The future of the emerging industry of WEEE recycling will be influenced critically by the events of the next decade. It is hoped that the contents of this monograph can provide guidance and information to those shaping these events. In recycling, as in life, we must focus on the future, for that is where all of us will spend the rest of our lives. This monograph has an ambition to contribute to this philosophy.

Košice, Slovakia, January 2015

Tomas Havlik
Head of Department of Non-ferrous Metals
and Waste Treatment
Technical University of Kosice
Slovak Republic

Contents

Abbreviations

2PC12	Oxime of 1-(2-pyridyl)-tridecane-1-one
2PC14	Oxime of 1-(2-pyridyl)-pentadecane-1-one
4PC12	Oxime of 1-(4-pyridyl)-tridecane-1-one
4PC14	Oxime of 1-(4-pyridyl)-pentadecane-1-one
ABS	Acrylonitrile-butadiene-styrene co-polymer
BAN	Basel Action Network
BioMinE	Biotechnologies for metal bearing materials in Europe
BTMHA	2-bromo-3,5,5-trimethylhexanoic acid
CCFLs	Cold cathode fluorescent lamps
CNT	Carbon nanotubes
CPU	Central Processing Unity
CRT	Cathode ray tubes
CYANEX 272	Bis(2,4,4-trimethylpentyl) phosphinic acid
CYANEX 301	Bis(2,4,4-trimethylpentyl)-dithiophosphinic acid
CYANEX 302	Bis(2,4,4-trimethylpentyl) monothiophosphinic acid
DEC	Diethyl carbonate
DEHPA	Di-2-ethylhexyl phosphoric acid
DIPSA	3,5-diisopropylsalicylic acid
DMC	Dimethyl carbonate
EC	Ethylene carbonate
EFSA	European Food Safety Authority
EMC	Ethyl methyl carbonate
EOL	End-of-life
EPA	Environmental Protection Agency (USA)
EPR	Extended Producer Responsibility
EPS	Exopolysaccharides
EU	European Unit
e-wastes	Electronic wastes
FDA	U.S. Food and Drug Administration
GNS	Grapheme nanosheets
HARL	Home Appliance Recycling Law (Japan)

HIPS	High impact polystyrene
HQ	8-hydroxyquinoline
HRJ 4277	Nonylsalicylic acid
IPS	In-Plane Switching
ITO	Indium oxide doped with tin oxide
LAGP	$Li_{1+x}Al_xGe_{2-x}(PO_4)_3$
LCA	Life Cycle Assessment
LCD	Liquid crystal display
LED	Light-emitting diode
LIBs	Lithium–Ion Batteries
LiPON	Lithium Phosphorous Oxynitride
LISICON	Li Super Ionic Conductor
LiTFSI	Lithium bis(trifluoromethansulfonyl)imide
LIX 984	1:1 of 5-nonylsalicylaldoxime and 2-hydroxy-5-nonylacetophenone oxime
LLT	$Li_{0.5-3x}La_{0.5+x}TiO_3$
LLZ	$Li_7La_3Zr_2O_{12}$
LTAP	$Li_{1+x}Ti_{2-x}Al_x(PO_4)_3$
LTO	$Li_4Ti_5O_{12}$
MCA	Multi Criteria Analysis
MFA	Material Flow Analysis
MOOS	Microorganisms
NASICON	Na Super Ionic Conductor
NiCd	Nickel Cadmium Batteries
NiMH	Nickel Metal Hydride Batteries
NMF	Non-metallic fraction
PAN	Poly(acrylonitrile)
PAHs	Polycyclic aromatic hydrocarbons
PBB	Polybrominated biphenyls
PBDE	Polybrominated diphenyl ethers
PC	Polycarbonate
PC 88A	2-ethylexyl hydrogen 2-ethylhexyl phosphonate
PCBs	Printed Circuit Boards
PCs	Personal Computers
PEO	Poly(ethylene oxide)
PGEs	Polymer gel electrolytes
PGM	Platinum Group Metals (Ru, Rh, Pd, Os, Ir, Pt)
PMMA	Poly(methyl methacrylate)
PP	Polypropylene
$PP_{24}TFSI$	N-butyl-N-ethylpiperidinium N,N-bis(trifluoromethane sulfonimide)
PPO/HIPS	Polyphenylene oxide/high impact polystyrene
PPS	Polysquarate
PVdF	Poly(vinylidene fluoride)
PVdF-HFP	Poly(vinylidene fluoride-hexafluoropropylene)

$Py_{24}TFSI$	N-n-butyl-N-ethylpyrrolidiniumN,N-bis(trifluoromethane) sulfonamide
$PYR_{14}TFSI$	N-methyl-N-butylpyrrolidinium bis(trifluoromethansulfonyl)imide
REE	Rare earth element
REM	Rare earth metal
REO	Rare earth oxide
RoHS	Restriction of Certain Hazardous Substances (Europe)
RTIL	Room temperature ionic liquids
S/L ratio	Solid/liquid ratio
SBR	Styrene-butadiene co-polymer
SCF	Supercritical fluid
SCWD	Supercritical water depolymerization
SCWO	Supercritical water oxidation
SEI	Solid electrolyte interface
SN	Succinonitrile
TBE	Tetrabromoethane
TBP	Tributylphosphine
TFT	Thin film transistors
TIBPS	Triisobutylphosphine sulphide
TOPS 99	Talcher Organo phosphorus solvent
TVs	Televisions
UN	United Nations
VOC	Volatile organic carbon
WEEE	Waste electrical and electronic equipment

Chapter 1
Introduction

Hugo Marcelo Veit and Andréa Moura Bernardes

Studies in the field of solid waste have grown a lot since the amount and rate of waste generation is increasing every year. It is therefore necessary to study alternatives to recycle and reuse solid waste, to extend the life cycle of materials, and to reduce the rate of extraction of natural resources to supply the industrial demand for raw materials.

One example is the consumer goods industry, especially the electronics industry, as the rapid development of this production chain, combined with the higher consumption of products by the population, generates an increasing number of defective and obsolete pieces of equipment, which need to be reused, recycled or correctly disposed.

The disposal of scrap and waste in landfills involves the loss of valuable materials and the increased risk of pollution by chemical compounds, for example. The zero waste generation concept and management systems that seeks to minimize, reuse, recycle, treat and dispose, should be inserted into all public and private policies.

Processes that reduces and reuse generated waste in a closed circuit are in everyone's interest. The foundations for an environmental oriented process, therefore, are:

(a) Primary Measures: Avoid or reduce the generation of waste;
(b) Secondary Measures: Reuse of waste;
(c) Tertiary Measures: Disposal with environmental compatibility, i.e., waste that can not be reused must be inert, so that it can be disposed without impacting adjacent areas.

H.M. Veit (✉) · A.M. Bernardes
Programa de Pós-Graduação em Engenharia de Minas, Metalúrgica e de Materiais (PPGE3M), Universidade Federal do Rio Grande do Sul (UFRGS), Porto Alegre, RS, Brazil
e-mail: hugo.veit@ufrgs.br

A.M. Bernardes
e-mail: amb@ufrgs.br

H.M. Veit and A. Moura Bernardes (eds.), *Electronic Waste*, Topics in Mining, Metallurgy and Materials Engineering, DOI 10.1007/978-3-319-15714-6_1

As such, two alternatives remain for the future of the consumer goods industry:

(a) Set up processes to reduce and reuse the generated waste in a closed loop;
(b) Produce a residue that can be viably inserted into the recycling industry.

The recycling of obsolete or defective materials is important because it represents an economic gain of the material itself, which is quite significant for metals like aluminum, lead, copper, and especially for noble metals: gold, silver, platinum and indium. Another advantage of reuse lies in energy savings. In the primary production of metals, the metal is obtained by ore reduction with high energy consumption. In the secondary process, the metal is obtained primarily from molten scrap, which is already in the metallic state, and the energy consumption is significantly lower.

It is important to characterize the products to define the recycling options. A characterization of waste and scrap consists of several stages and has as main objective the determination of the physico-chemical properties of the materials and the evaluation of hazardousness. In fact, it is important to determine the chemical composition and physical properties of the materials to define the recycling possibilities, enabling their use as secondary raw material for different industries. In order to evaluate the environmental compatibility of waste and scrap that will be used as raw material or that is to be disposed in landfills, different characteristics must be evaluated. In general, a characterization of waste and scrap should include, among other characteristics, the evaluation of the items listed below:

- Heterogeneity.
- Chemical composition, considering total elemental composition, concentration of salts, oil and grease content, pH, redox potential, etc.
- Physical characteristics: density, particle size, permeability, porosity, humidity etc.
- Hazardousness characteristics: corrosivity, reactivity, radioactivity, toxicity, etc. Regarding toxicity, leaching and solubilization tests can assist in the waste characterization.

The recycling conditions can only be established after a complete characterization of the waste and scrap, i.e., only after thoroughly understanding the waste, should the recycling processes and routes be chosen and applied.

This book presents an overview of the characterization of electronic waste. In addition, processing techniques for the recovery of metals, polymers and ceramics are described. Considering the contents described here, this work can serve as a source of information and as an educational technical reference for practicing scientists and engineers, as well as for students.

Chapter 2
Electronic Waste: Generation and Management

Hugo Marcelo Veit and Andréa Moura Bernardes

2.1 Electronic Waste in the World

The electronics industry is one of the most important industries in the world. It has grown steadily in recent decades, generates a great number of jobs, promotes technological development and, at the same time, fuels a high demand for raw materials that are considered scarce or rare (e.g. precious metals and rare earths elements).

This development affects the environment in two ways: first through the large and growing amount of equipment that is discarded annually and second through the extraction of natural raw materials to supply the demand of the new equipment industry. Both can be measured by the amount of equipment that is produced and discarded annually by many countries.

The main manufacturers and generators of electronic waste are considered to be developed or developing countries. According to Robinson [1], most electronic waste is generated in Europe and the United States. However, China, the countries of Eastern Europe and Latin America are becoming large generators of electronic waste.

Estimates indicate that about 40 million tonnes of this waste are generated each year, or approximately 5 % of all solid waste generated in the world [2]. Another study, from Greenpeace, estimated that globally 20–50 million tonnes of WEEE (waste electrical and electronic equipment) are discarded annually, with Asian countries disposing up to 12 million tonnes [3].

Robinson [1] also estimated that computers, mobile phones and television sets alone would correspond to 5.5 million tonnes of electronic waste generated on a

H.M. Veit (✉) · A.M. Bernardes
Programa de Pós-Graduação em Engenharia de Minas, Metalúrgica e de Materiais—PPGE3M, Universidade Federal do Rio Grande do Sul—UFRGS, Porto Alegre, RS, Brazil
e-mail: hugo.veit@ufrgs.br

A.M. Bernardes
e-mail: amb@ufrgs.br

H.M. Veit and A. Moura Bernardes (eds.), *Electronic Waste*, Topics in Mining, Metallurgy and Materials Engineering, DOI 10.1007/978-3-319-15714-6_2

global scale in 2010. Moreover, for 2015, this figure was projected to be 9.8 million tonnes. According to the author, by volume electronic waste may correspond to about 8 % of all municipal waste generated.

For monitors there is no specific number on the amounts discarded annually. However, for some countries, like the USA, China and Taiwan, it is estimated that 3.2 million, 5 million and 1 million televisions and monitors, respectively, are disposed of annually [1].

Similarly, in relation to computers, there is no exact number for discarded equipments, but this number may be assumed to be proportional to the quantity of computers in use worldwide. According to the Computer Industry Almanac [4], the number of computers in use worldwide in the 80s was 4.8 million. In 2000 this number had increased to 553 million and for 2015 a number of 2,020–2,070 million has been projected. The 5 countries with the largest number of computers in use until the end of 2011 were USA (310.6 million), China (195.1), Japan (98.1), Germany (71.5) and India (57).

World data indicate that, in 2011, 372 million PCs were sold worldwide and sales are projected to hit 517 million units by 2015. The three countries with the highest sales in 2011 were China, USA and Brazil [4].

The life cycle of many electronic goods has been substantially shortened due to advancements in electronics, attractive consumer designs, marketing and compatibility issues. For example, the average life cycle of a new computer has decreased from 4.5 years in 1992 to an estimated 2 years in 2005 and is further decreasing [5].

According to the Basel Action Network (BAN), an organization that tracks the flow of toxic waste in the world, eight in ten unused computers in the USA end up in Asian countries, such as India and China, where recycling costs are lower. Africa has also become a center for the receipt of electronic waste [6].

Data from the United Nations (UN) for 2005 quantified the per capita generation of electronic waste in developing countries. According to this study, Brazil and Mexico are the developing countries that generate the most electronic waste from computers, with about 0.5 and 0.45 kg/person/year, respectively, ahead of such countries as South Africa, with 0.4 kg/person/year, and China, with 0.2 kg/person/year [7].

In addition to computers, mobile phones are another type of electrical and electronic equipment with great production and sales.

In the USA, an EPA survey showed that approximately 125 million mobile phones are discarded annually, which means 65,000 tonnes of electronic waste. The same survey showed that a mobile phone has an average life cycle of 9–18 months [8].

Another study has shown that the global number of obsolete mobile phones is estimated to be more than 500 million and continues to grow rapidly [9]. In November 2013, the number of mobile phones in use in Brazil reached 271 million, i.e., well above the Brazilian population, which is 200 million [10].

As far as the recycling of mobile phone devices is concerned, there are no concrete data on recycling rates in the world, but research conducted by Nokia in 13 countries showed that only 3 % of mobile phones are recycled [11].

Currently, the lion's share of electronic waste comes from computers, mobile phones, televisions, monitors, cathode ray tubes (CRT) and printers [12]. However, new technologies such as liquid crystal display (LCD) are now replacing older devices, generating new types of electronic waste.

In developing countries like China and India, low cost and easy operation result in most electronic waste being further treated in backyards or small workshops using primary methods such as open burning and acid wash, especially in order to recover metals of economic interest [13].

At the same time, large companies like Umicore, Noranda, Boliden, and Cimélia, among others, have published papers on the process of refining metals, mainly in order to recover precious metals. These industries process various types of industrial wastes and by-products of non-ferrous metals, including Printed Circuit Boards (PCBs), recovering about 20 precious metals and other non-ferrous metals. The processes used follow a complex flowchart, with several steps that include pyrometallurgical, hydrometallurgical and electrochemical techniques and operations [14–17]. Considering that information and communication technologies continue to spread throughout the world, recycling technologies must be well managed and established worldwide.

2.2 The Problem of WEEE

Several environmental protection agencies around the world consider WEEE to be hazardous waste because they have chemicals compounds in their composition that are toxic and harmful to human health and to the environment [18, 19].

The chemical composition of WEEE varies according to each product. For example, LED TVs have a higher amount of polymer, while stoves and microwaves have a larger amount of metals. In fact, the chemical composition depends on several factors, such as the type of WEEE, year of manufacture, manufacturer's brand and country of origin. Usually, WEEE are classified by type, using the regulations of environmental agencies, such as Directive 2002/96/EC of the European Union [20].

In general, a mixture of metals can be found in WEEE, such as copper, iron, aluminum, brass and even precious metals, such as gold, silver and palladium, in addition to a mixture of polymers, such as polyethylene, polypropylene, polyurethane and others. WEEE may also include ceramic materials, such as glass, and other inorganic, organic and even radioactive materials [1].

WEEE is considered toxic to human health and to the environment because it often has inorganic compounds in its composition, such as mercury, lead, cadmium, nickel, antimony, arsenic and chromium, in addition to such organic compounds as polychlorinated biphenyls, chlorofluorocarbons, polycyclic aromatic hydrocarbons and polyhalogenated, among others [1].

The concentration of these compounds and substances in WEEE varies with the product type, year of manufacture and the manufacturer itself. For example, old refrigerators contained chlorofluorocarbon refrigerant, but with the development of the Montreal Protocol in 1989, and its ratification by member countries, the manufacturers of refrigerators replaced this toxic gas by another inert gas in the new refrigerator models [21].

Some efforts have been made to prevent/reduce the use of toxic products in electronics equipment. The most important was, probably, Directive 2002/95/EC of the European Union, also known as RoHS—Restriction of Certain Hazardous Substances [22]. This resolution banned the manufacture and sale of consumer electronics in the European Union that containing lead, mercury, cadmium, hexavalent chromium, polybrominated biphenyls and/or polybrominated diphenyl ethers. As such, many countries that have trade relations with the EU were obliged to manufacture products without these substances.

Despite these efforts, WEEE disposal is still a worldwide challenge, since in many countries there is no structured system of reverse logistics and most WEEE is still disposed in landfills or in the open places exposed to the weather [23–26]. This form of disposal can contaminate the soil or subsoil with metals and/or toxic substances from WEEE. Some studies, such as those by the EPA [27] and Spalvins [28], show that contamination by WEEE in landfills is difficult to quantify because there are many external factors affecting the decomposition of chemicals, including temperature, pressure, pH, and oxygenation of the medium, among others. In another EPA study [29], it is pointed out that 2–5 % from all solid waste sent to landfills in the USA comes from electronic waste.

The incineration of WEEE is an alternative to landfill, but can also cause environmental problems. Many components of WEEE have organic compounds in the composition that, when incinerated, can generate dioxins and furans [30]. The presence of halogens in WEEE can be explained by the addition of flame-retardants containing bromine. Approximately 12.5 % of all types of WEEE contain halogenated compounds [31]. If the burning of WEEE is performed without proper environmental precautions, the release of polybrominated polyphenyl compounds, and others like it, will occur. The primary means of preventing the release of dioxins and furans during the incineration of equipment is a gas treatment system. However, these are very expensive pieces of equipment for the companies that perform these services. As a result, many companies are burning without an adequate treatment of gases.

EPA studies [32, 33] regarding the disposal of WEEE in the USA showed that, in 2001, about 2 million tonnes of WEEE were generated, of which approximately 90 % was sent to landfills. In 2009, 2.37 million tonnes were generated, but the percentage of recycling reached 25 %, i.e., much higher than in 2001.

The recycling of WEEE is a growing trend around the world, but for a successful recycling process, the strategies and technologies used for the collection and processing of e-waste are very important. And although many countries already legislation in place dealing with e-waste, especially the countries of the European Union, Japan and the USA, the management of this waste is still deficient in the majority of countries around the world [34].

2.3 WEEE Management

Extensive research is currently under way into e-waste management in order to mitigate problems at both the national and international levels. Several tools have been developed and applied to e-waste management, including: LCA (Life Cycle Assessment), MFA (Material Flow Analysis), MCA (Multi Criteria Analysis) and EPR (Extended Producer Responsibility) [35].

These management tools, combined with the existing laws in different countries, can help improve the disposal of electronic waste in the world, increasing the reuse of materials and reducing environmental impacts.

In Europe, a number of legislative documents have been drafted and/or implemented requiring manufacturers and other stakeholders to adopt an environmental approach to design and assess the environmental impact of their products throughout their lifecycle [36]. The main laws and regulations are: the WEEE (Waste Electrical and Electronic Equipment) and the RoHS (Restriction of the Use of Certain Hazardous Substances) Directives, created by the European Union. The WEEE Directive makes manufacturers responsible for collecting, recycling and/or disposing of waste from electrical and electronic equipment. The RoHS Directive, which came into force in July 2006, prohibits the use of six hazardous substances in the manufacture of products, including electronics equipment. These substances are cadmium (Cd), mercury (Hg), lead (Pb), hexavalent chromium (Cr VI), polybrominated biphenyls (PBB) and polybrominated diphenyl ethers (PBDE) [37].

In Asia, the rapid economic growth in most countries is increasing the amounts of WEEE generated. In addition, Asian countries are the recipient of most of the WEEE generated globally. Ketai et al. [38] estimate that about 4K tonnes of WEEE are discarded globally per hour, 80 % of which is exported to Asia, with China receiving 90 % of this share. In March 2007, China has adopted legislation to control the pollution caused by electronic products, focusing on the restrictions of using hazardous substances in electronic equipment (unofficially referred to as China RoHS) [26]. In Japan, in order to decrease the environmental impact of WEEE, the Home Appliance Recycling Law (HARL) was introduced and adopted in April 2001 [39]. The law was enacted for four main sources of consumer WEEE in Japan, namely: refrigerators, washing machines, TVs and air conditioning units. In 2008, the law was revised to incorporate LCD and plasma TVsn and clothes dryers.

The basic principles of Japan's HARL are the replacement of toxic substances and the increase of recyclability, encouraging recycling and prohibiting inappropriate deposits. Through this law, the consumer pays a fee to recycle the product, with the government being responsible for the collection and for the reverse logistics system. The manufacturers are responsible for the recycling of materials and the proper reuse of toxic components, i.e., the responsibilities for the waste are shared. However, the HARL has been criticized for excluding certain types of WEEE, such as mobile phones. There is no recycling law that specifically targets mobile phone waste, despite the fact that more than 25 million units are discarded every year. It is estimated that only 40 % of mobile phone waste is collected by retailers [40].

As far as Africa is concerned, it is known that many African countries receive second-hand equipment. Most electronic equipment exports to Africa are not pre-tested for functionality. Consequently, it is not possible to assess whether these exports are legally defined as hazardous waste under the Basel Convention [41]. The Basel Convention was adopted on March 22, 1989 and came into force on May 5, 1992. It deals with the control of transboundary movements of hazardous waste and its deposits. This Convention provides that hazardous waste must be disposed of in the country of origin. The Base Convention considers electronic waste as hazardous and its export could be allowed only under special conditions [41].

Except for South Africa, where an increase in material recovery activities has been reported, data on the recycling of WEEE in Africa is scarce [42]. Despite working under generally deplorable social and environmental conditions, the informal collectors, dismantlers and recyclers in Africa are playing an increasingly important role in the processing of WEEE [43, 44].

Sales of electronic equipment in Latin/South American countries have increased rapidly, approaching the level of industrialized countries, with increasing WEEE generation as a result [45]. However, WEEE recycling is mostly restricted to disassembly companies. Although a number of traditional metal recycling companies in countries such as Chile, Argentina, Peru, Colombia and Brazil had discovered the WEEE recycling markets, processed quantities are still on a modest level [45]. Several countries are engaged in policy formulation related to WEEE, with Brazil seemingly the frontrunner in this issue [46]. In Brazil, the National Policy on Solid Waste enacted in August 2010 requires manufacturers, importers, distributors and retailers of electronics products and its components to structure and implement reverse logistics systems to collect the product after its use by consumers [47].

In North America, the USA is the main generator of electronic waste with approximately 7.5 kg/person/year [26]. Several states in the United States have implemented WEEE programs, but specific legislation is rare. Initiatives to adopt legislation on the federal level have not been successful so far [48, 49]. There are currently at least 20 states and one municipality (New York City) that have legislation about WEEE. In Canada, the situation is similar to the USA, i.e., there is no national legislation concerning electronic waste, but several provinces already have their own legislation [50].

The disposal of all this waste generated globally is extremely worrying. The best way to manage this waste is through the recycling of the materials present in it. The chemical composition of WEEE is extremely heterogeneous, varying according to the type of equipment, year of manufacture, manufacturer, and country of origin, among other factors. In fact, polymers, ceramics and metals can be recovered from electronic waste. The different metals present draw the greatest attentions because of their economic value. These metals may be present in different types of components, in variables amounts, pure or as alloys.

Printed circuit boards (PCBs), present in all types of electronics equipments, are of major interest because they are considered secondary raw materials that are rich in copper and precious metals such as gold, silver and palladium. For example, a single mobile phone can contain high concentrations of gold (24 mg), silver (250 mg) and palladium (9 mg) [51]. When compared to the average contents of the primary sources (minerals), these values reveal a secondary source of high metal concentration. This stimulates the recovery of these metals, ensuring metallic resources for future use and avoiding all environmental impacts related to their primary extraction [51].

Table 2.1 shows the concentration of metals found by several authors in the PCBs from mobile phones. One can see that the values are not entirely similar, due to the reasons mentioned above.

Considering this metal content, the management of WEEE for its recycling is increasing in every country. E-waste management is different in developed and developing countries. Overall, there are two types of facilities in developed countries engaged in the recycling chain, according to the nature of the methods involved. The first group contains the facilities that are principally engaged in the dismantling and mechanical processing of e-waste for the recovery of raw materials. The second group uses metallurgical processes to recover metals. In contrast, the e-waste recycling sector in developing countries is largely unregulated, and e-waste is often processed to recover valuable materials in small workshops using rudimentary recycling methods. The informal e-waste recycling prevalent in developing countries is associated with severe environmental pollution and occupational exposure to the e-waste-derived chemicals [58]. Although the costs of recycling can be smaller in backyards common to the developing countries, the use of large-scale processes enables the recovery of much higher yields of valuable substances, such as precious metals, than backyard operations. For relevant parts such as printed circuit boards, this higher yield usually overcomes the cost disadvantages [13].

Table 2.1 Metal concentration (wt%) of printed circuit boards reported in different studies

wt%	Sum [52]	Guo et al. [53]	Yang et al. [54]	Park and Fray [55]	Yamane et al. [56]	Tuncuk et al. [57]
Gold	0.1	0.008	–	0.025	0.00	0.035
Silver	0.2	0.33	–	0.100	0.21	0.138
Copper	20	26.8	25.06	16.0	34.49	13
Nickel	2	0.47	0.0024	1	2.63	0.1
Tin	4	1.0	–	3.0	3.39	0.5

References

1. Robinson BH (2009) E-waste: an assessment of global production and environmental impacts. Sci Total Environ 408:183–191
2. Huisman J, Magalini F, Kuehr R et al (2008) Review of directive 2002/96 on waste electrical and electronic equipment (WEEE). United Nations University, Bonn, 2007
3. Greenpeace—The e-waste problem—Greenpeace International. http://www.greenpeace.org/international/campaigns/toxics/electronics/the-e-waste-problem#. Accessed 11 Jun 2014
4. Computer Industry Almanac. http://www.c-i-a.com. Accessed 10 Jun 2014
5. Widmer R, Oswald-Krapf H, Sinha-Khetriwal D et al (2005) Global perspectives on e-waste. Environ Impact Assess Rev 25:436–458
6. Basel Action Network (Ban). http://www.ban.org. Accessed 10 Jun 2014
7. UNEP (2012) United Nations Environment Programme. http://www.unep.org. Accessed 11 Apr 2014
8. EPA (United States Environmental Protection Agency) (2004) The life cycle of a mobile phone—solid waste and emergency response
9. Skerlos SJ, Basdere B (2003) Environmental and economic view on cellular telephone remanufacturing. In: proceedings colloquium e-ecological manufacturing. Uni-Edition, Berlim, pp 143–148
10. Anatel—Agência Nacional de Telecomunicações. http://www.anatel.gov.br. Accessed 23 Jan 2014
11. Nokia Brasil (2008) Reciclagem de telefones celulares. http://www.nokia.com.br/a-nokia/meio-ambiente/we-recicle/por-que-reciclar. Accessed 05 Mar 2014
12. Wong MH (2007) Export of toxic chemicals—a review of the case of uncontrolled electronic-waste recycling. Environ Pollution 149:131–140
13. Sepúlveda A, Schluep M, Renaud FG et al (2010) A review of the environmental fate and effects of hazardous substances released from electrical and electronic equipments during recycling: examples from China and India. Environ Impact Assess Rev 30:28–41
14. Hageluken C (2007) Recycling of e-scrap in a global environment: opportunities and challenges. In: Rajeshwari KV, Basu S, Johri R (eds) Tackling e-waste towards efficient management techniques. TERI Press, New Delhi, pp 87–104
15. Brusselaers J (2005) An eco-efficient solution for plastics-metals-mixtures from electronic waste: the integrated metals smelter. In: 5th Identiplast 2005. The Biennial conference on the recycling and recovery of plastics identifying the opportunities for plastics recovery, Brussels, Belgium
16. Heukelem A (2004) Eco efficient optimization of pre-processing and metal smelting. In: Electronics goes green. Berlin, pp 657–662, 6–8 Sept 2004
17. Hageluken C (2006) Recycling of electronic scrap at umicore's integrated metals smelter and refinery, World of Metallurgy. Erzmetall 59(3):152–161
18. Jinhui L, Brenda NLN, Lili L, Nana Z, Keli Y, Lixia Z (2013) Regional or global WEEE recycling. Where to go? Waste Manage 33:923–934
19. Rafia A, Muhammad MM, Rulia A, Jarita BD (2013) Survey and analysis of public knowledge, awareness and willingness to pay in Kuala Lumpur, Malaysia e a case study on household WEEE management. J Clean Prod 52:185–193
20. Directiva 2002/96/CE—European Community. http://eur-lex.europa.eu/resource.html?uri=cellar:ac89e64f-a4a5-4c13-8d96-1fd1d6bcaa49.0010.02/DOC_1&format=PDF. Accessed 24 Feb 2013
21. United Nations Environment Programme (UNEP)—Montreal Protocol. http://ozone.unep.org/new_site/en/montreal_protocol.php. Accessed 29 Jan 2013
22. Directiva 2002/95/CE—European Community http://eur-lex.europa.eu/LexUriServ/LexUriServ.do?uri=OJ:L:2003:037:0019:0023:PT:PDF. Accessed 29 Jan 2013

23. Elshkaki A (2005) Dynamic stock modeling: a method for the identification and estimation of future waste streams and emissions based on past production and product stock characteristics. Energy 30:1353–1363
24. Bernardes AM, Tenório JAS, Espinosa DCR (2003) Collection and recycling of portable batteries: a worldwide overview compared to the brazilian situation. J Power Sources 124:586–592
25. Masahiro O, Hirofumi S, Atsushi T (2013) Toxic metals in WEEE: characterization and substance flow analysis in waste treatment processes. Sci Total Environ 463–464:1124–1132
26. Ongondo FO, Williams ID, Cherrett TJ (2011) How are WEEE doing? A global review of the management of electrical and electronic wastes. Waste Manage 31:714–730
27. Epa Ireland (2001) Waste from electrical and electronic equipment. http://www.environ.ie/en/Publications/Environment/Waste/WEEE/FileDownLoad,1341,en.pdf. Accessed 30 Jan 2013
28. Spalvins E (2008) Impact of electronic waste disposal on lead concentrations in landfill Leachate. Environ Sci Technol 42:7452–7458
29. Global Futures Foundation (GFF) (2001) Computers, e-waste, and product stewardship: is California ready for the challenge? Report for the U. S. Environmental Protection Agency, San Francisco, CA. http://infohouse.p2ric.org/ref/41/40164.htm. Accessed 11 Jun 2014
30. Hall W, Willians P (2007) Analysis of products from the pyrolysis of plastics recovered from the commercial scale recycling of waste electrical and electronic equipment. J Anal Appl Pyrolysis 79:375–386
31. Vehlow J Recycling of bromine from plastics containing brominated flame retardants in state-of the-art combustion facilities. http://www.bsef.org/china/uploads/library/Tamara.pdf. Accessed 30 Jan 2014
32. Environmental Protection Agency (2001) Municipal solid waste in the United States: 2001 facts and figures. Appendix C, computer electronics in municipal solid waste. http://www.epa.gov/osw/nonhaz/municipal/pubs/msw2001.pdf. Accessed 05 Feb 2014
33. Environmental Protection Agency (2011) Electronics waste management in the United States through 2009, office of resource conservation and recovery. http://www.epa.gov/waste/conserve/materials/ecycling/docs/fullbaselinereport2011.pdf. Accessed 23 Feb 2014
34. Oliveira CR, Bernardes AM, Gerbase AE (2012) Collection and recycling of electronic scrap: a worldwide overview and comparison with the Brazilian situation. Waste Manage 32:1592–1610
35. Kiddee P, Naidu R, Wong MH (2013) Electronic waste management approaches: an overview. Waste Manage 33:1237–1250
36. Goosey M (2004) End-of-life electronics legislation—an industry perspective. Circuit World 30(2):41–45
37. European Commission Environment. http://ec.europa.eu/environment/waste/weee/index_en.htm. Accessed 05 Dec 2013
38. Ketai H, Li L, Wenying D (2008) Research on recovery logistics network of waste electronic and electrical equipment in China. In: Industrial electronics and applications ICIEA. 3rd IEEE conference, pp 1797–1802
39. Zhang B, Kimura F (2006) Network based evaluation framework for EEE to comply with environment regulations. In: Mechatronics and automation, proceedings of the 2006 IEEE international conference, pp 797–802
40. Hosoda E (2007) International aspects of recycling of electrical and electronic equipment: material circulation in the East Asian region. J Mater Cycles Waste Manage 9(2):140–150
41. Basel Convention. http://www.basel.int/. Accessed 05 Dec 2013
42. Nnorom IC, Osibanjo O (2008) Electronic waste (e-waste): material flows and management practices in Nigeria. Waste Manage 28(8):1472–1479
43. Dittke S, Newson G, Kane C et al (2008) A material recovery facility in Cape Town, South Africa, as a replicable concept for sustainable e-waste management and recycling in developing countries. In: Global symposium on recycling, waste treatment and clean technology, Cancun, Mexico, 12–15 Oct. Available from: http://ewasteguide.info/2008_Schluep_REWAS

44. Rochat D, Laissaoui SE (2008) Technical report on the assessment of E-waste management in Morocco. EMPA—Swiss federal laboratories for materials testing and research, Switzerland. Available from: http://ewasteguide.info/Laissaoui_2008_CMPP. Accessed 11 Jun 2014
45. Silva U, Ott D, Boeni H (2008) E-waste recycling in Latin America: overview, challenges and potential. In: Global symposium on recycling, waste treatment and clean technology, Cancun, Mexico, 12–15 Oct
46. Horne RE, Gertsakis J (2006) A literature review on the environmental and health impacts of waste electrical and electronic equipment—2. International policy and regulation [ministry for the environment]. RMIT University (Centre for Design), New Zealand. Available from: http://www.mfe.govt.nz/publications/waste/weee-literature-review-jun06/html/page3.html. Accessed 11 Jun 2014
47. República Federativa do Brasil. Lei 12.305, de 2 de agosto de 2010. Política Nacional de Resíduos Sólidos. Available from: http://www.planalto.gov.br/ccivil_03/_ato2007-2010/2010/lei/l12305.htm. Accessed 11 Jun 2014
48. UNEP and UNU (United Nations Environment Programme and United Nations University) (2009) Sustainable innovation and technology transfer industrial sector studies—recycling from e-waste to resources. Berlin, Germany
49. Bohr P (2007) The economics of electronics recycling: new approaches to extended producer responsibility. PhD thesis, Technischen Universität, Berlin, Germany
50. Environment Canada (2003) Baseline study of end-of-life electrical and electronic equipment in Canada. Available from: http://www.epsc.ca/. Accessed 11 Jun 2014
51. Kim E (2011) Selective recovery of gold from waste mobile phone PCBs by hydrometallurgical process. J Hazard Mater 198:206–215
52. Sum EYL (1991) The recovery of metals from electronic scrap. JOM 43(4):53–61
53. Guo J, Guo J, Xu Z (2009) Recycling of non-metallic fractions from waste printed circuit boards: a review. J Hazard Mater 168(2–3):567–590
54. Yang T (2009) Factors influencing bioleaching copper from waste printed circuit boards by Acidithiobacillus ferrooxidans. Hydrometallurgy 97(1–2):29–32
55. Park YJ, Derek JF (2009) Recovery of high purity precious metals from printed circuit boards. J Hazard Mater 164(2–3):1152–1158
56. Yamane LH, Moraes VT, Espinosa DCR, Tenório JAS (2011) Recycling of WEEE: characterization of spent printed circuit boards from mobile phones and computers. Waste Manage 31:2553–2558
57. Tuncuk A, Stazi V, Akcil A, Yazici EY, Deveci H (2012) Aqueous metal recovery techniques from e-scrap: hydrometallurgy in recycling. Miner Eng 25(1):28–37
58. Tsydenova O, Bengtsson M (2011) Chemical hazards associated with treatment of waste electrical and electronic equipment. Waste Manage 31:45–58

Chapter 3
Processing Techniques

Hugo Marcelo Veit

Electronic waste processing is very complex due to the great heterogeneity of its composition and its poor compatibility with the environment.

The first step is usually manual disassembly, where certain components (casings, external cables, CRTs, PCBs, batteries, etc.) are separated. Following disassembly, the technologies used for the treatment and recycling of electronic waste include mechanical, chemical and thermal processes. For metals recovery there are four main routes: Mechanical Processing, Hydrometallurgy, Electrometallurgy and Pyrometallurgy. There are also references in the literature of the use of biotechnology, or a combination of various techniques. The following chapter presents a brief summary of each route. From Chap. 6 onwards, each one of these routes will be discussed individually.

3.1 Mechanical Processing

The mechanical processing of WEEE is used to select and separate materials and the separation is based on steps of mineral processing techniques. As far as scraps are concerned, mechanical processing is generally seen as a pretreatment for the separation of materials, and associates with different separation stages of WEEE components [1–8]. Currently, several studies have been conducted for the mechanical processing of electronic scrap, which contains a mixture of materials. Comminution, classification and separation [1, 6] (by density difference, particle size, magnetic and electrical properties) are therefore part of this process. Figure 3.1 presents these methods. The metal fraction obtained after the

H.M. Veit (✉)
Programa de Pós-Graduação em Engenharia de Minas, Metalúrgica e de Materiais (PPGE3M), Universidade Federal do Rio Grande do Sul (UFRGS), Porto Alegre, RS, Brazil
e-mail: hugo.veit@ufrgs.br

H.M. Veit and A. Moura Bernardes (eds.), *Electronic Waste*, Topics in Mining, Metallurgy and Materials Engineering, DOI 10.1007/978-3-319-15714-6_3

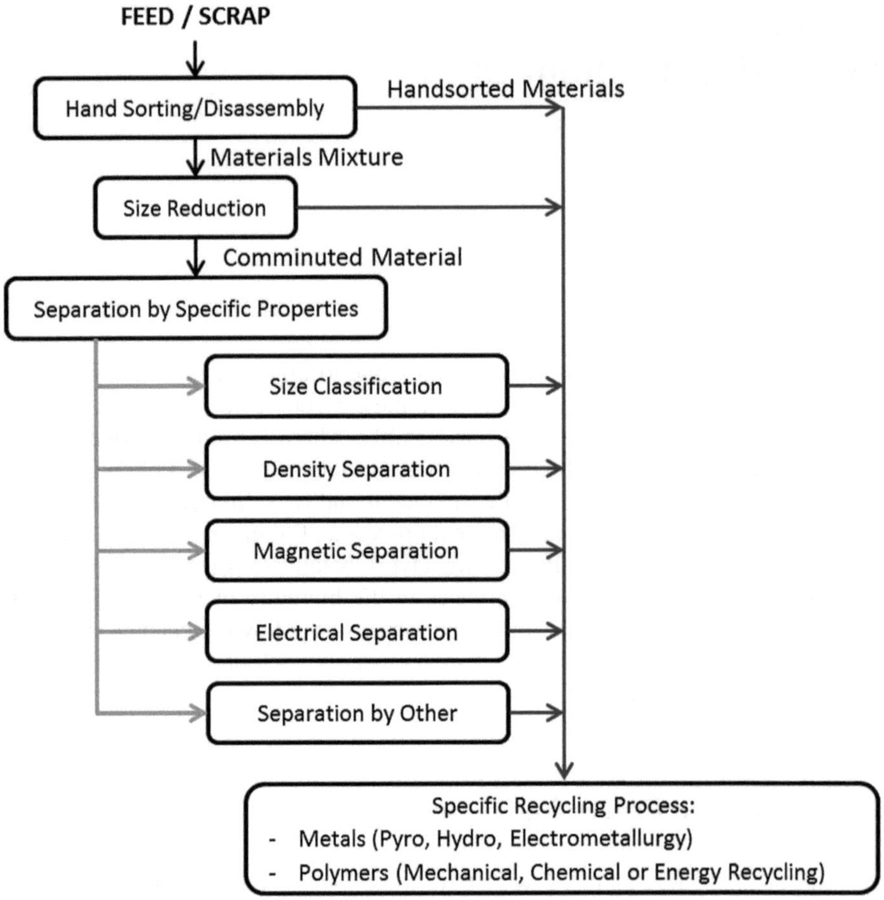

Fig. 3.1 Material selection and separation methods based on mineral processing techniques

mechanical processing step is sent to hydro-, electro- and/or pyrometallurgical processes. The ceramic and polymer fractions are sent to specific routes and this subject will be discussed in Chap. 9.

3.2 Hidrometallurgy

The initial steps in hydrometallurgical processing consist of a number of acidic or caustic attacks to dissolve the solid material [9, 10]. In the following steps the solutions are subjected to separation processes such as solvent extraction, precipitation, cementation, ion exchange, filtration and distillation to isolate and concentrate the metals of interest [11].

The main advantages of hydrometallurgical processing of electronic waste, when compared to pyrometallurgical methods, are [10, 11]:

- Reduced risk of air pollution
- Higher selectivity to metals
- Lower process costs (e.g. low power consumption and reuse of chemical reagents).

The disadvantages are:

- Difficulty in processing more complex electronic scraps
- Need for mechanical pre-treatment of the scrap to reduce volume
- The chemical dissolution is effective only if the metal is exposed
- Large volume of solutions
- The wastewater can be corrosive, toxic or both
- Generation of solid waste.

3.3 Biotechnology

The use of bacteria in metals recycling has been described in the literature [12, 13], specifically in the dissolution of metals and the recovery of gold from electronic waste. This process is still quite restricted in terms of large scale implementation, but several studies are being conducted that should be able to produce processes with low operating costs and low investment in equipment, in addition to generating little waste, effluents or toxic gases.

Biohydrometallurgical processes include the leaching of metals by different bacterial cultures, as reported in the studies by Sum [14] and Brandl et al. [12]. In the experiments described by Sum [14], the electronic waste was treated with a solution containing 10 g L^{-1} of Fe^{+3} and a culture of the bacterium (YTL-2) at pH < 2.5 and a temperature of 20–35 °C. After 2 days, it was possible to recover 97 % of the gold, in addition to regenerating the bacteria and reusing the solution. In the study by Brandl et al. [12], microbiological processes were applied to electronic waste. They found that bacteria (*Thiobacillus thiooxidans*, *ferroxidans*) and fungus (*Aspergillus niger*, *Penicillium simplicissimum*) grow in the presence of this type of residue. In this study, high quantities of metals, such as Cu, Ni, Zn and Al, were leached.

The main limitations of biohydrometallurgical processes are the long periods necessary for the leaching and the need of the metal to be exposed, i.e., the metals content must be mainly located on the surface layer.

3.4 Electrometallurgy

Most electrometallurgical processes associated with the recycling of electronic waste are steps of the electrowinning process that ultimately seeks to recover a pure metal [11]. Electrochemical processes are usually performed in aqueous electrolytes or molten salts and can be used to recover metals from various types of waste.

Metal concentrates obtained by hydrometallurgical processes (e.g., selective dissolution, ion exchange or solvent extraction) can be electrodeposited from aqueous solutions on the cathode [14]. The advantages of electrometallurgical processes are:

- Few steps
- Higher selectivity for desired metals
- The electrolyte can be reused
- Pure metals can be obtained.

The main limitation is the need of a pre-treatment (usually based on mechanical and hydrometallurgical processes).

3.5 Pirometallurgy

Pyrometallurgical processes, notably smelting, has become a traditional method to recover metals from e-waste [15–17].

The conventional pyrometallurgical processing mechanism consists essentially of concentrating metals in a metallic phase and rejecting most other materials in a slag and/or gas phase [11].

Pyrometallurgical processing has some advantages, such as: applicability to any type of electronic waste, no need for pre-treatment and few steps in the process. Some of the methods involving thermal processing of electronic waste, however, can cause the following problems [14, 18]:

- Polymers and other insulating materials become a source of air pollution through the formation of dioxins and furans.
- Some metals can be lost through volatilization of their chlorides.
- Ceramic and glass components present in the scrap increase the amount of slag in the furnace, increasing the losses of precious and base metals.
- Recovery of some metals is low (e.g. Sn and Pb) or almost impossible (e.g. Al and Zn).

References

1. Hayes PC (1993) Process principles in minerals and materials production. Hayes Publishing Co. Brisbane, p 29
2. Zhang S, Forssberg E (1999) Intelligent liberation and classification of electronic scrap. Powder Technol 105:295–301
3. Tenório JAS, Menetti RP, Chaves AP (1997) Production of non-ferrous metallic concentrates from electronic scrap. TMS Ann Meet. Orlando, EUA, pp 505–509
4. Zhang S, Forssberg E (1997) Mechanical separation-oriented characterization of electronic scrap. Resour Conserv Recycl 21:247–269

5. Distin PA (1995) Metals recovery from electronic scrap using mineral processing methods. In: 34th annual conference of metallurgists of CIM. Vancouver, Canada, pp 245–252
6. Wills BA (1988) Mineral processing technology, 4th edn. Pergamon Press, Great Britain
7. Veit HM, Pereira CC, Bernardes AM (2002) Using mechanical processing in recycling printed wiring boards. JOM 54(6):45–47
8. Veit HM, Pereira CC, Hörlle LF, Bernardes AM (2002) Recycling of metals from PCB by dense medium separation processes. In: Recycling and waste treatment in mineral and metal processing: technical and economic aspects. Lulea, pp 16–20
9. Pozzo RL, Malicsi AS, Iwasaki I (1991) Removal of lead from printed circuit board scrap by an electrodissolution-delamination method. Resour Conserv Recycl 5:21–34
10. Gloe K, Mühl P, Knothe M (1990) Recovery of precious metals from electronic scrap, in particular from waste products of the thick-layer technique. Hydrometallurgy 25:99–110
11. Hoffmann JE (1992) Recovering precious metals from electronic scrap. JOM 44(7):43–48
12. Brandl H, Bosshard R, Wegmann M (2001) Computer-munching microbes: metal leaching from electronic scrap by bacteria and fungi. Hydrometallurgy 59:319–326
13. Nakazawa H, Shouming W, Kudo Y (2002) Bioleaching of waste printed wiring board using thiobacillus ferrooxidans. In: Recycling and waste treatment in mineral and metal processing: technical and economic aspects. Lulea, pp 16–20
14. Sum EYL (1991) The recovery of metals from electronic scrap. JOM, Abr, pp 53–61
15. Cui J, Zhang L (2008) Metallurgical recovery of metals from electronic waste: a review. J Hazard Mater 158:228–256
16. Tsydenova O, Bengtsson M (2011) Chemical hazards associated with treatment of waste electrical and electronic equipment. Waste Manag 31:45–58
17. Oliveira CR, Bernardes AM, Gerbase AE (2012) Collection and recycling of electronic scrap: a worldwide overview and comparison with the Brazilian situation. Waste Manag 32:1592–1610
18. Felix N, Riet CV (1994) Recycling of electronic scrap at um's hoboken smelter. In: International precious metals conference. Vancouver, Canada, pp 159–169

Chapter 4
Mechanical Processing

Angela Cristina Kasper, Nichele Cristina de Freitas Juchneski and Hugo Marcelo Veit

Mechanical processing includes the steps of comminution, size classification, gravity separation, magnetic and electrostatic separation, among others. Each of these techniques is detailed below.

4.1 Comminution

Size reduction, or comminution, or grinding are synonymous and represent the breaking of the material through brute force by the movement of "hammers", "balls" or "knives" in a closed compartment. It is one of the oldest unit operations for the reduction of solid materials, in addition to being an important step in many processes where the raw material is converted into intermediate or final products [1].

In the recycling industry, the comminution processes has a similar objective as those of the mineral, chemical and pharmaceutical industries, i.e., to release a valuable component that is encapsulated in the particles, producing particles of desired size and shape, increasing the surface area of the particles and generating particles with more active surface [2]. In the case of electronic waste, comminution processes are intended to achieve a suitable liberation degree of the different metals that compose it, thus allowing the subsequent use of separation processes. In fact, although degree of liberation was defined by Gaudin [3] to be applied to ores, the same definition may be satisfactorily used for waste recycling [2].

A.C. Kasper (✉) · H.M. Veit
Programa de Pós-Graduação em Engenharia de Minas, Metalúrgica e de Materiais (PPGE3M), Universidade Federal do Rio Grande do Sul (UFRGS), Porto Alegre, RS, Brazil
e-mail: angelakasper@globo.com

H.M. Veit
e-mail: hugo.veit@ufrgs.br

N.C.F. Juchneski
Nuclear Industries of Brazil—INB, Caetite, Brazil
e-mail: ni.juchneski@gmail.com

H.M. Veit and A. Moura Bernardes (eds.), *Electronic Waste*, Topics in Mining, Metallurgy and Materials Engineering, DOI 10.1007/978-3-319-15714-6_4

In short, the definition states that the degree of liberation is the percentage of a phase or material that occurs as free particles in relation to the total sample. On the other hand, the entrapment degree is the percentage of the material that occurs as trapped/encapsulated particles with respect to the total.

Zhang and Forssberg [4] studied the liberation degree of metals versus non-metals in the recycling of printed circuit boards. The liberation degree, in this case, is extremely important for the success of subsequent mechanical processing. The researchers noted that the metals present in the electronic waste can be easily liberated when the comminution method reaches particles smaller than 2.0 mm, achieving virtually total liberation. However, it is important to evaluate the energy required to achieve this level of comminution. According to Masuda et al. [1] comminution energy efficiency is very low, and the energy required is inversely proportional to the size of the particles produced. Jäckel et al. [5] conducted a study on various types of mills, evaluating the milling time and energy consumption for various types of materials.

The comminution process occurs by applying a force in different forms: pressure, impact, abrasion or cutting. The fragmentation of the material occurs mainly by the movement of different devices, which may be balls, knives or hammers. Two types of mills are used in solid waste processing: vertical and horizontal mills. In vertical mills, the residue is inserted at the top. To be fragmented, it has to be hit as fast as possible by the moving "hammer" or "knives". Particle size control is adjusted by the distance between the "hammer" or "knives" and walls. In horizontal mills, "hammers" oscillate above a grid that can be changed depending on the size of the required product [6].

Equipments commonly used for the comminution of WEEE are swing-hammer type shredders and knife mills.

The size reduction of scrap from electronic appliances is generally carried out by swing-hammer shredders. These shredders have become important for the comminution of materials with ductile behavior, such as scrap from household appliances and electrical devices. This is due to the complex stressing modes occurring inside this type of equipment, which are well suited for the comminution of thin-walled metal scraps [7, 8].

In a Hammer Mill, the size reduction is achieved by hammers capable of rotating around an axis. The material is destroyed by collision with the hammers and the walls of the mill. A hammer mill has a grid or screen at the bottom through which the material falls after being sufficiently reduced [9]. The product size can be adjusted by changing the grid or screens size, and by lengthening or shortening of the arms of the hammer. Figure 4.1 shows an example of a hammer mill.

In a Knife Mill, on the other hand, the size reduction is achieved by cutting the material using knives mounted on two shafts rotating in opposite directions. There is often a grid in the back. Knife mills are especially suitable to reduce elastic materials (ductile) [6]. Figure 4.2 shows a model of knife mill.

Cryogenic Grinding is another comminution process where the material is reduced in size by freezing at very low temperatures with the aid of liquid nitrogen. The increased fragility of the material improves the efficiency of the mill and therefore the liberation degree of materials.

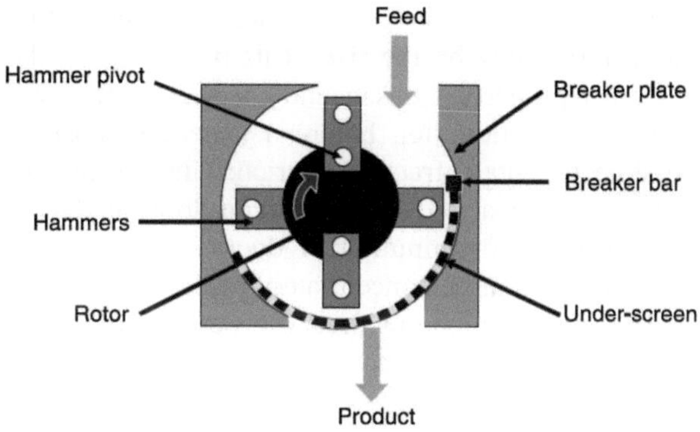

Fig. 4.1 Hammer mill [10]

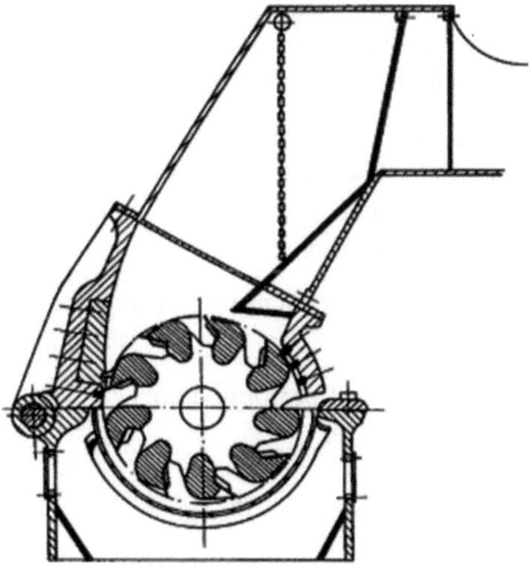

Fig. 4.2 Knife mill [11]

Each material may therefore receive special types of treatment in the comminution of electronic waste because of the wide variety of materials found. The scraps may be processed only once or several times by the same mill or by a combination of the different available mills.

4.2 Size Separation

After passing through the comminution phase, the material generally has to be classified. Size classification is based on the separation of particles according to their different size. One of the simplest and most widely used forms for size separation is sieving. This is done with sieves that consist of perforated surfaces over which the particles are moved. The smaller particles pass through those perforations while the larger particles remain on the sieve. This process separates the material only by the size of its particles, and the material is not identified by any other property. Consequently, sieves are frequently used in materials recovery as a classification step before a materials separation process. Sometimes, it is possible to obtain fractions enriched in certain materials after this step, which can then be separated from the main flow [6]. This separation is a function of the difference in comminution of ductile or brittle materials, which facilitates the formation of a more concentrated fraction of a particular material.

The main equipment used for sieving are vibrating sieves, rotary sieves and vibration machines with air injection [1, 12]. In vibrating sieves, which are the most common, a perpendicular vibration is applied to the sieve surface. This equipment can be applied to both fine and coarse particles [1].

According to Kasper [13], the sieving efficiency can be affected by factors related to the material, equipment characteristics and the type of movement applied to the sieve.

The main factors concerning the material are:

- Density
- Particle Size Distribution
- Particle Shapes
- Particle Surfaces

The key factors concerning the equipment are:

- Sieve area
- Percentage of perforated area
- Hole diameter
- Hole shapes
- Sieve thickness

The factors concerning the movement applied to the sieve to be considered are:

- Frequency
- Amplitude and direction of vibration
- Tilt angle
- Processing time

4.3 Density Separation

The separation by density difference, also known as gravimetric separation, is an old process. The separation of gold using this principle, for example, dates back to at least 3000 BC, according to the writings of ancient Egypt [14]. It is one of the oldest forms of mineral processing, but its mechanisms are not yet fully understood. The main mechanisms in the process are: differential acceleration, sedimentation rates, differential speed in laminar flow, interstitial consolidation and the effect of shear forces [15].

Density separation is a physical process that is based on the difference between the rates of sedimentation or terminal velocity of particles, and its principle is the separation of the materials by different densities, sizes and shapes [14, 15].

This technique was originally used in the mineral processing industry and in the cleaning of coal, and today it has also been studied in solid waste recycling processes. This process can be used on electronic waste consisting essentially of polymers with densities below 2.0 g/cm^3, light metals, mainly aluminum, with a density of 2.7 g/cm^3, and heavy metals, predominantly Cu, Fe, Ni and Pb, with a density greater than 7 g/cm^3 [16]. Regardless of these differences in the density of materials, when the materials of interest are aggregated into one component, it will be necessary to release them, i.e., a comminution step will be required to liberate the components of composites and assemblies. Since specific materials are released, it is feasible to use various density separation systems. Among the different processes studied, air classifiers have been used to separate non-metallic components and the dense medium and concentrators processes have been applied for fractions of non-ferrous metals [17].

The main processes known for density separation are: dense medium separation, suspensions, jigs, air and flowing film concentrators.

4.3.1 Separation by Dense Medium

The separation by dense medium consists in introducing a quantity of a sample of known mass into a container containing a liquid of known density, subsequently separating the floating and sunk material. The sunken fraction may pass through successive stages of the same procedure. This technique can be used for various purposes, such as gravity concentration tests, process control, design and applicability tests of density separation units, tests to determine mineral composition of the product, evaluation of the efficiency of other separation processes, analysis of the release of material, among others [18].

The dense medium used in the assay depends on the density of the material and may consist of dissolved salts, such as calcium chloride in water, with a density of approximately 1.50 g/cm^3, or zinc chloride, with 1.75 g/cm^3. These are low density media and have been used, for example, for the separation of coal. Heavy organic liquids are proposed for industrial use, but because of their toxicity and costs, they are limited to small-scale use.

In addition to the choice of a medium with the proper density, the viscosity should also be taken into consideration. It should be low enough to allow the particles to separate quickly [14].

Table 4.1 shows some dense liquids as well as their respective densities.

Table 4.1 Dense liquids

Dense liquid	Density (g/cm^3)
H_2O	1.0
$NaCl + H_2O$	1.2
$ZnCl_2$	1.75
$CaCl_2 + H_2O$	1.5
Bromoform	2.89
Acetone: TBE = 0.27:1	2.5
TBE	2.96
Clerici solution	4.28

4.3.2 Separation via Suspensions

Separation via suspension is a process based on the density difference of the products of interest. The product is placed in a liquid or pulp with an intermediate density and the fraction with a higher density than the medium sinks, while the fraction with a lower density than the medium floats.

Industrially, there are three types of processes: salt solutions, suspensions and autogenous media. In the separation of coal, for example, salts such as calcium chloride dissolved in water were used. However, due to its high costs, this operation is no longer used on a large scale.

A suspension can be defined as a liquid that keeps insoluble solids in a state of fluidization. In the case of dense medium separation, the suspension is a heterogeneous system consisting of a insoluble solid dispersed in water, which behaves like a liquid, such as magnetite and FeSi [15, 19].

The autogenous medium used in the water-only-cyclones corresponds to a mixture of water with the material of interest. In suspensions, a finely comminuted solid is mixed with water in order to obtain a pulp with a defined density, which rises by the increasing of the amount of solids added to the pulp.

The equipments which use suspensions to separate materials are static separators and cyclones.

In static separators, the driving force for the separation of the floated fraction from the sunken fraction is gravity. These devices are tanks that operate continuously, which means that ore or material and the dense medium are fed continuously [19].

Static separators have a high separation efficiency and the performance is little influenced by particle size when compared to other types of separators, such as hydraulic and pneumatic separators. However, due to the agitation and the height of the separator, finer particles and particles with densities close to the dense medium, often do not have enough time to sink or float. This behavior is associated with a low sedimentation rate caused by the high viscosity and yield stress of the medium. therefore, by reducing the particle size of the feed, there is a reduction in separation efficiency.

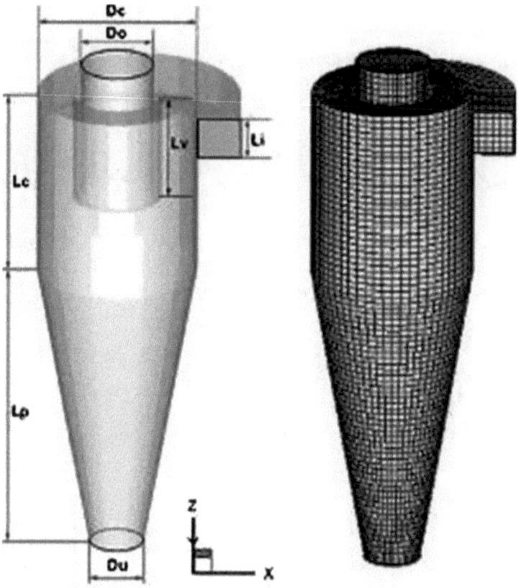

Fig. 4.3 Dense medium cyclone [20]

By using cyclones, the high centrifugal force improves the separation of finer particles when compared with other methods, making this a very useful piece of equipment. Figure 4.3 shows a dense medium cyclone.

4.3.3 *Jigs*

Concentration in jigs is one of the most used processes in gravity concentration. The extensive use of jigs is mainly a result of their low operating cost, in addition to their capacity to treat fractions of a wide variety of sizes in a single unit [21]. The jig is a hydraulic separation process, which consists of repeated dilation followed by vertical compaction of a bed of particles by water movement. The result is the separation of the particles in layers where density increases from top to bottom.

The stratification into a bed of particles is achieved through a repeated pulse of a fluid stream sent upward through the bed. The particle beds are expanded so that when the pulse ceases, the particles are under the influence of gravity. Figure 4.4 illustrates the compaction and expansion of the bed with the heavier particles.

The process of expansion and compaction of the bed is repeated in a cyclic operation until the light and heavy particles have been stratified according to their specific gravity. The pulse frequency varies typically from 50 to 300 cycles per minute [14].

Figure 4.5 shows the schematic drawing of a jig.

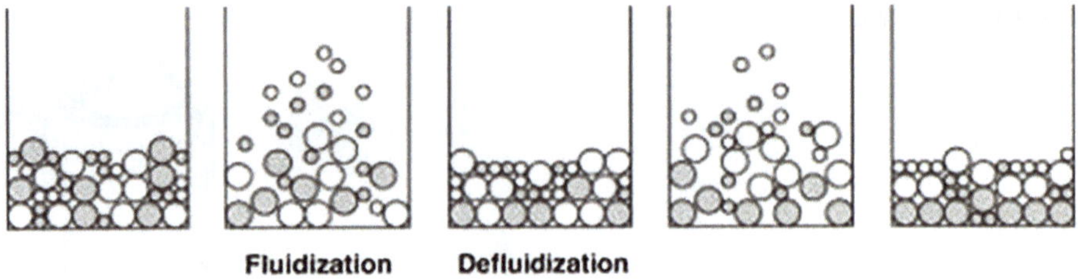

Fig. 4.4 Compaction and expansion of a bed [22]

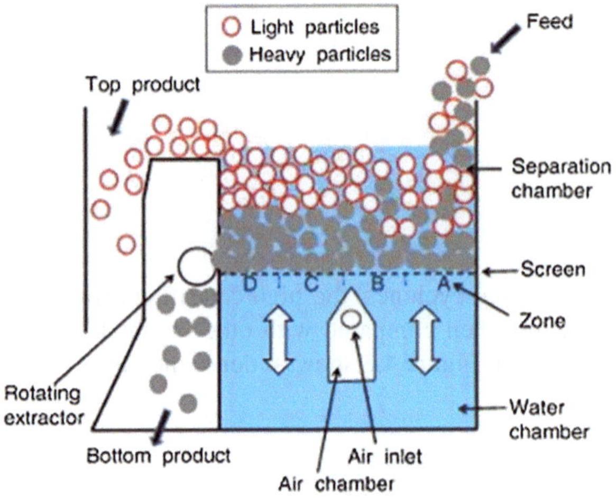

Fig. 4.5 Schematic drawing of a jig [14, 23]

4.3.4 Flowing Film Concentrators

The separation using flowing film concentrators uses two main methods: table and spiral concentrators.

A typical concentrating table consists of a wooden deck coated with a material with a high friction coefficient (rubber or plastic), partially covered with riffles. The differential movement of the riffled deck and the flowing water are used to separate particles. The vibratory motion is asymmetric, being slow in the forward direction and fast in the reverse direction. This differential movement carries the particles that are in contact with the deck table, by friction, in the direction of the motion.

The Wilfley table was designed for the concentration of ore and, from the beginning, it has been widely used for this purpose. When the table is equipped with decks designed for coal washing, it is known as a Massco table. Since the introduction of the Wilfley table, many different brands of table have been developed for use with minerals and coal [14]. In general, it is considered to be the most efficient equipment for treating materials with fine grain sizes. However, due

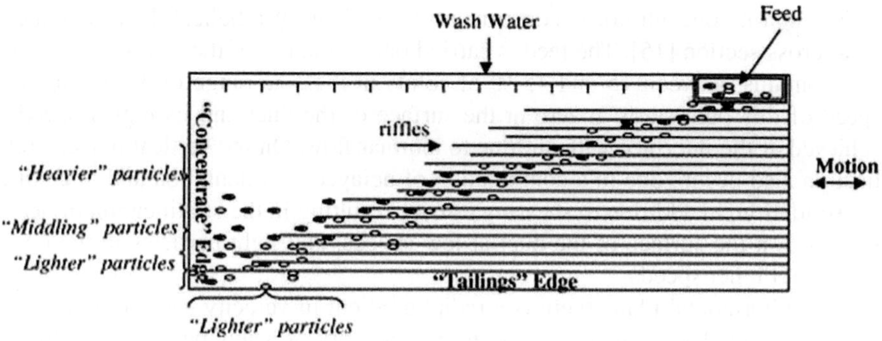

Fig. 4.6 Shaking table [24]

to its low processing capacity (<2 tons/h), it is a device that is mainly used for the cleaning of primary or secondary concentrated ore and free gold alluvial ores [15].

The table consists of a gently sloping flat bed, with a number of riffles along the direction of motion. The riffles are inclined to the opposite end of the oscillating unit. The feed is approximately 25 % solids (by weight), and with the agitation, the particles are spread on the table.

Wash water is introduced along the upper edge of the deck to assist in the separation and transport of particles on the table. As a result, the particles move diagonally across the platform from the feed end [14]. Figure 4.6 shows a schematic drawing of a table.

The driving separation mechanisms in concentrating tables are diverse and vary according to the region of the deck. In the region of the table with riffles, the sample, fed transversely to the riffles, undergoes the effect of the asymmetrical movement of the table, which results in forward movement of the particles, and the fine and heavy particles moving over the coarse and light. In the spaces between the riffles, the particles are separated due to the expansion caused by the asymmetric movement of the table and the turbulence of the pulp through the riffles. The bed between the riffles behaves like a small jig with delayed sedimentation and interstitial consolidation, causing the heavy and fine material to stay closer to the surface than the large and light material.

The upper layers are carried over the riffles by the new feeding and by the film washing with a transverse flow. The riffles, decrease in height along their length so that, progressively, fine and heavy particles are brought into contact with the washing water film that passes over the riffles. The result of the asymmetrical movement towards the riffles and the differential speed of a perpendicular laminar flow is the scattering of material [15].

Other flowing film concentrators are the spiral ones. Spirals and static sluice-type gravity concentrators have been used since 1943 to restore and clean anthracite fine grain [21]. This equipment is another variant of gravity separation through density differences and centrifugal force.

The spiral concentrator is constructed in the form of a helical duct of semicircular cross section [15]. The feed is carried out, at the top of the spiral, with a pulp that contains between 15 and 45 % of solids in the size range of 3–75 mm. The speed of this pulp starts at zero at the surface of the duct and its higher speed is achieved at the interface with air, due to laminar flow. On the vertical plane, stratification also occurs due to a combination of delayed sedimentation and interstitial consolidation, in addition to shearing efforts, resulting in the stratification of heavy particles on the surface of the duct at low speeds, and light particles in the upper stream at higher speeds.

In the horizontal plane there is a radial gradient in velocity caused by the helical path, which has a smaller effect on the trajectory of heavy particles. For lighter particles, however, this effect is substantial and due to the centrifugal force they tend towards an outer trajectory [15].

The particles with higher density are collected through the door located at the lowest point of the cross section. The washing water, added to the inner edge of the current, flows to the opposite side of the concentrate. Adjustable splitters are placed to control the width of the collected concentrate.

4.3.5 Air Separation

Air separation is a unit operation used for the separation of solid dispersed particles based on differences in size and density [25].

Air separation equipment, also known as dry tables, is used in situations where water is scarce, or when water-soluble salts are present in the sample, or when wetting the sample is not desirable. The basic principles of dry tables do not differ from wet tables [26].

The main advantages are the low operation and investment costs and the high capacity. In addition, the process does not generate effluents. This process has some limitations, however, such as poor accuracy and high sensitivity to moisture.

The pneumatic tables work by means of a constant movement of air through a porous bed to move the product fed along a flat riffled of smooth platform. The results of particle separation are obtained based on differences in size and density.

The feed is held at one end of the table where there is an upward airflow. The oscillations of the table and airflow allow the stratification of the bed. As a result, the heavier particles are directed to the base, while the light particles are directed to the top [15].

Zhang et al. [27] showed that the particles in the bed stratify according to their densities so that light particles tend to move up while heavier particles move down. If the particles to be processed are different in size and density, the stratification becomes complicated. For example, thin copper wires (heavier particles) stratify with polymers (small particles) in the recovery of copper from electronic waste [28].

4.4 Magnetic Separation

Magnetic separation is a method widely used in the separation and/or concentration of materials, both in the production of minerals and in the recycling of waste from electrical and electronic equipment and others types of scrap. Magnetic separation can be regarded as an environmentally friendly technique [29].

Magnetic separation is governed by two factors: the behavior of particles of different materials when exposed to the same magnetic field and the magnetic forces acting on them. The responses of different particles to the applied field result in the attraction or repulsion of the same field [30].

Magnetic susceptibility is the property that determines the type of response of a material to an external magnetic field, i.e., the magnitude of the interaction between this material and an external magnetic field. This magnetic susceptibility can be described as [6]:

$$X = M/H \tag{4.1}$$

where: X is the magnetic susceptibility, M is the intensity of magnetization of the material and H is the applied magnetic field.

The intensity of the field within the material subjected to a magnetic field is given by Eq. 4.2 [30]:

$$B = \mu 0(H + M) \quad \text{(Tesla)} \tag{4.2}$$

where: B is the field strength and $\mu 0$ is the magnetic vacuum permeability. Tesla (T) $= 10^4$ Gauss, which corresponds to 1 $N\,A^{-1}\,m^{-1}$.

The materials can therefore be classified according to their magnetic susceptibility, or according to their behavior in an external magnetic field, as diamagnetic materials and paramagnetic materials. The ferromagnetic materials can also be included in the category of paramagnetic materials. However, their response to the external magnetic field make them particularly interesting for the application of magnetic separation techniques [6].

Diamagnetic materials: are materials where the induced magnetic field opposes the external magnetic field, generating a repulsion of low intensity. That is, these materials are repelled by the external magnetic field. Examples: copper, mercury, gold, silver, zinc, lead, among others [31].

Paramagnetic materials: are materials that are characterized by the fact that they have atomic magnetic dipoles with a tendency to align themselves parallel to an external magnetic field, causing them to be weakly attracted to the magnets. Examples: aluminum, sodium, manganese, tin, chromium, platinum, palladium, among others [31].

Ferromagnetic materials: are materials that exhibit strong interactions between their atomic magnetic dipole, which produces an internal alignment in the same direction as the applied magnetic field. These materials undergo a major influence from the external magnetic field, being strongly attracted by magnets. Examples: iron, cobalt, nickel, among others [31].

In diamagnetic and paramagnetic materials, the relationship between M and H are approximately linear. That is, X is constant for a given material. The diamagnetic materials have a very small negative value for X, while paramagnetic materials present a small positive susceptibility. Ferromagnetic materials have a large X value that varies depending on the applied magnetic field and the magnetic history of the sample [6].

All materials are therefore affected in some way by the presence of a magnetic field. So when a particle susceptible to a magnetic field is placed in a magnetic separator, the magnetic force will act on the particle, enabling its separation from other remaining materials. The magnetic characteristics of a material depend on such factors as the elements constituting the material, their states of ionization and their concentrations in the solid, the crystal structure of the solid and, in some cases, on the previous history of the material (in case of ferromagnetic materials) [6].

The separation is possible by the use of techniques such as high intensity magnetic separation (HIMS) and high gradient magnetic separation (HGMS), using electromagnets (low conductivity), superconductors or permanent magnets; by the development of magnetic materials, for example, with rare earth; and by the introduction of the ferromagnetic matrix, that enable an increase in the gradient in the high intensity field. All these developments have brought significant progress in the use of the magnetic separation technique [32]. These progresses have made possible the separation of both strongly magnetic and weakly material, even when they are in the form of finely dispersed particles. The selection of the class of separator to be used, therefore, depends on several considerations. The most important is the size distribution and the magnetic distribution of the product being processed and also the capacity of the equipment.

In modern equipment, both the field and the gradient (variation of field strength that results in the force acting on the material, causing the attraction or repulsion) are responsible for the separation process.

There are several types of magnetic separators that can be classified in many different ways. However, the most common classification classifies them according to their use in dry and wet separators. They can also be subdivided according to the characteristics of the induction field (separators of low and high intensities). The shape of the elements that perform the work of separating the equipment has significant influence on the classification of the separators. As such, they are called drum roller, induced cross belt, and carousel separators, among others [32].

The most widely used magnetic separators are described below [32, 33].

4.4.1 Dry Separators

Low intensity (0.06–0.1 T): The main applications of this type of separator are removing iron and strongly magnetic impurities. or concentrating a valuable and strongly magnetic component (for example, nickel). It is used in the separation or concentration of scrap material, ferromagnetic minerals and roasted products. Examples of equipments are: suspended magnets, magnetic pulleys and magnetic drums.

High intensity (0.1–1.0 T): This type was developed because of the need for processing coarse and weakly magnetic materials. Equipment using rare earth magnets allow the generation of a magnetic field on the rolls. Although magnetic fields can not be easily varied, a careful selection of the permanent magnet material and the optimization of the geometrical configuration of the roll makes it possible to process materials of different size ranges and different magnetic susceptibilities. This method is used in the concentration or separation of scrap materials and paramagnetic minerals. Examples of equipments are: cross belts, discs and rolls.

High gradient (1.0–5.5 T): This is used in the separation or concentration of scrap materials and paramagnetic minerals, coal, kaolin and water treatment. Examples of equipments are: rare earth element rolls and open gradient separators.

4.4.2 Wet Separators

Low intensity (0.06–0.1 T): These pieces of equipment are provided in two types of configurations, radial or axial. In a radial configuration, the polarity of the permanent magnets alternates across the drum's width, while in an axial arrangement, the poles alternate along the circumference. The radial configuration is typically used in applications where the quantity of strongly magnetic material to be recovered is more important. The axial configuration is preferred when the quality of the magnetic product is more important. It is used for the separation or concentration of scrap material, ferromagnetic minerals and roasted products. Examples of equipments are: submerged belt and wet drums separators.

High gradient (1.0–5.5 T): Developed by the introduction of a matrix in a magnetic circuit, these separators enable magnetic separation to be applied to materials that previously were considered too weakly magnetic. It is used for the separation or concentration of scrap materials and paramagnetic minerals, coal, kaolin and water treatment. Examples of equipments are: matrices and stationary and movable drum.

The control of the intensity of the magnetic field allows for the selective separation of particles with different values in magnetic susceptibility. At low field strengths, minerals with high susceptibility are separated, while at high intensities, materials with lower susceptibility are separated. In general, the control of the field strength is done by varying the electrical current of electromagnets. For some equipments, one can vary the field by prior adjustment of the distance between the poles. Equipment with electromagnets have mechanisms that soften and relax the variation of field strength, while the equipment with permanent magnets do not have these mechanisms [32].

In general, the control of the transport of particles through the magnetic field and the amount of material passing through the magnetic field, can be considered as ways of improving the selectivity of separation.

The performance of the magnetic separators, such as the high gradient magnetic separator, may be affected by a number of variables, such as the magnetic field intensity, the type and form of the material, and the orientation of the magnetic field, among others [34].

The technological advances of recent years have made it possible to use the technique of magnetic separation on a wider range of applications. It can be used, for example, to remove particulate ferrous flows in excess of 100 m^3/h, frequently found in process waters from the steel processing industry [35]; to remove phosphate from sewage [36]; to separate particles of different sizes in liquid ferrofluids [37]; and to clean up soil contaminated with lead and other metals [38].

In the mining industry, magnetic separation is used for the concentration of hematite, for kaolin beneficiation; for the removal of magnetic impurities; for the concentration of the cassiterite, scheelite, chromite, feldspar and quartz; for the removal of magnetite from asbestos and phosphate ores; for the purifying of talc; for the recovery of wolframite and molybdenum from sulphide minerals contained in the tailings of beneficiations; and for the beneficiation of uranium and heavy minerals (ilmenite, rutile) [32].

In recycling processes, magnetic separation is widely used to separate ferrous scraps of the automotive industry in general, in addition to being used for the treatment of automotive shredder residue [39]. Magnetic separation is also used as a step of separation and recovery of materials from electronic scrap and batteries [9, 12, 40–45].

4.5 Electrostatic Separation

The electrostatic separation is a separation or concentration method of materials based on the difference in electrical conductivity (resistivity) between the components of a mixture of particles under the action of an electric field.

The origin of the term electrostatic separation is in fact associated to the first purely electrostatic separators that were developed, without the so-called ionic flow. However, currently devices are used in which the power is applied in the form of ionic flow [46].

Two conditions are basically needed for electrostatic separation: an electric field of sufficient intensity to separate an electrically charged particle, and particles with a superficial electrical charge or subject to an induced polarization, which allows them to undergo the influence of the electric field. The electrification by induction occurs when a not charged particle is brought into contact with a grounded conductive surface (rotor) in the presence of an electric field, assuming the same polarity and the same potential of any surface. The electrification in this case depends on the field strength and on the nature of the particles [46].

Through induction, both the conductive and the dielectric material acquire an electric charge, the latter acquiring polarization due to the transfer of charges. As such, the conductive particles quickly assume the potential of the rotor (opposite to the discharge electrode) and are attracted to the electrode, while the non-conductive particles, the ones that are incapable of acquiring a charge, are attracted to the rotor or remain polarized, passing through the field largely intact and allowing the separation of materials with conductive particles from materials with non-conductive particles [32].

The particles can acquire a surface charge primarily by three mechanisms: attrition (Triboelectric separation), ion bombardment (corona effect), or eddy current.

4.5.1 Electrification by Contact or Friction (Triboelectrostatic Separation)

The electrification by contact or friction uses the difference in the electronic structure of the surface of the particles in the charging mechanism. The electrification occurs by transfer of electrons (charges), which occurs when two particles of different materials are rubbed against each other, resulting in positively charged particles and negatively charged particles. Under influence of an electric field, these charged particles move toward the oppositely polarized electrodes.

The magnitude of the electrification is the result of two processes, the charge transfer that occurs during contact, and the backflow of charge that occurs when the materials are separated. Repeated contacts of the weakly conductive electrical particles result in the creation of a "static" charge. Although commonly observed, electrical contact is difficult to control because of the variability of material properties [47].

Triboelectrostatic separation enables the separation of different types of plastics according to the difference between their electrical properties. This has some advantages, such as independence of particle shape, low power consumption and high performance [12].

4.5.2 Electrification by Ion Bombardment (Corona Separation)

In the electrification by ion bombardment, a very high voltage difference between the electrode and the rotor generates a charge transfer between both, i.e., an ion flux and charged atoms or molecules are generated by the gas. The high voltage electrostatic field is generated by a corona electrode and an additional electrode. The particles of metals and non-metals that enter this area are subject, at the same time, to the electrostatic induction and ion bombardment (corona charging). The metal particles are discharged quickly to the electrode and as a result they fall into the deposit faster. In addition, the charged particles of non-metals are attracted by the rotor and remain "trapped" for a longer time before falling into the deposits [46].

The corona electrostatic separators usually consist of: a grounded rotating drum and two electrodes (one electrode and a static Corona electrode), a cleaning brush and various particle trajectories. The mixture of particles is fed onto the surface of the rotating drum, where the ion bombardment is applied through the ionization electrode. The particles under the effect of the corona are negatively charged and remain attached to the surface of the drum until they penetrate the region of action of the static electrode. The function of the static electrode is to revert, by induction, the charges of the conductive particles, causing them to detach from the drum surface by changing their path, allowing their collection in the box of conductive material. The non-conductive particles remain negatively charged and, therefore, glued to the surface of the drum until they are removed with the brush and the alternating current electrode.

Corona electrostatic separation is a suitable technique for fine particles with sizes ranging from 0.1 to 5.0 mm. However, the most suitable particle size is between 0.6 and 1.2 mm [48].

The electrostatic separation technique using a corona separator is frequently used for the separation of metals (liable to induction) from non-metals on printed circuit boards, enabling the concentration of these metals.

4.5.3 Eddy Current (Foucault Current)

This method is based on the induction of Foucault currents (parasite currents) in conductive particles of non-ferrous materials to create a magnetic field around the particle. The polarity of this field is the same as that of the magnet, causing the material to be repelled, i.e., these particles are deflected while the non-conducting particles are not deflected, creating two streams of material, allowing their separation. This technique makes it possible to separate ferrous from non-ferrous metals, or to remove non-ferrous materials [49].

The technique was initially developed to recover non-ferrous metals of vehicle scraps and to treat municipal solid waste, but now it's being widely employed for other purposes.

In general, the efficiency of electrostatic separation depends on such factors as: the mechanism system that produces the surface charges to be separated; the particle release; the absence of dust and humidity, which affect the conductivity of air and, consequently, the charging conductivity of the material and the material itself; and the rotating velocity of the electrostatic separator [50–52].

The main variables of the equipment that may affect the separation are: electrode position, defined by two parameters, the distance and angular location of the rotor; the position of the divisors of recovered products; the voltage used in the electrode; the speed of the rotor; and the atmospheric pressure [48].

The efficiency of this type of separation is influenced by many factors, such as: material properties; the level of high voltage; the electrode configuration; the feed rate; the particle size; the roll speed; and the environmental conditions [53].

Before electrostatic separation is applied, it is essential that the material is submitted to a grinding process, because one of the prerequisites for efficient material separation is an appropriate particle size. For large particles the release between metals and non-metals is poor. This occurs because these particles have a low surface charge due to the low specific surface area and the electrostatic force acting on them is therefore smaller than the individual weight of each particle, which decreases the efficiency of the process. This problem can be solved by repeated grinding processes. However, excessive grinding is almost unavoidable, even when the grinding process is operated under optimal conditions. The non-conductive fine particles (<0.045 mm) have a large specific area that increases the electrostatic force acting upon them. The presence of this type of particle affects the efficiency of the separation process, mainly by: being susceptible to the formation of aggregates, reducing the stability of the process, accumulating on the surface of the electrode during the process [54].

Studies show that levels exceeding 10 % of such particles significantly affect the electrostatic separation process. An increase in the content of these particles can lead to an increase of 200 % of the quantity of product present in the mixed fraction (fraction containing a mixture of metals and non-metals) and a reduction of up to 20 % in the quantity of product present in the conductive portion (metals) [54]. The particle size considered ideal for the electrostatic separation is between 0.6 and 1.2 mm [51].

The new settings of corona electrostatic separators with two-rolls produce higher separation efficiencies than the classic settings that used only one corona roll. This occurs because the new configuration is more compact and efficient, and unlike what one might think, it is not a simple connection of two cylinders. These new devices make use of gravity to move the grain mixture for the second step, eliminating a transport unit, thereby reducing costs and processing time. In these devices, each roller has a function [55, 56]. In this new configuration, the amount of product in the conductive fraction (metallic) increased by 34 %, while the amount of product in the mixed fraction was reduced by 73 %. The amount of metal present in the mixed fraction was reduced to 33 % compared to 58 % in the classical hardware configuration, which indicates a more efficient process. Furthermore, spark discharges and the impact of small particles on the electrostatic electrode were reduced by improving the stability and reliability of the separation process [55].

The optimization of the process parameters can be achieved through the use of software to simulate the electric field distribution, evaluating how variations in the parameters of the electrodes and applied voltage influencing the electric field distribution will affect the separation. The comparison of the simulated results with experimental data enables the optimization of operating parameters, improving the results of separation [57].

References

1. Masuda H, Higashitani K, Yoshida H (eds) (2007) Powder technology: handling and operations, process instrumentation, and working hazards. Taylor and Francis Group, EUA, UK, 412p
2. Schubert W, Tomas J (2007) Liberation of valuables embedded in particle compounds and solid waste. Handb Powder Technol 12:989–1017
3. Gaudin AM (1939) Principles of mineral dressing. McGraw Hill Book Co., New York, pp 70–71
4. Zhang S, Forssberg E (1999) Intelligent Liberation and classification of electronic scrap. Powder Technol 105:295–301
5. Jackel HG, Schubert G, Sander S (2002) Size reduction of scrap by means of horizontal shaft shredders of the swing hammer type. In: Recycling and waste treatment in mineral and metal processing: technical and economic aspects. Lulea, Sweden, 16–20 June 2002
6. Hayes PC (1993) Process principles in minerals and materials production. Hayes Publishing CO., Brisbane, p 29
7. Russo P, Aboussouan L, Birat JP, Sander S, Jäckel HG, Schubert G (2004) Mechanical and metallurgical study of the fragmentation of end-of-life goods in a scrap shredder. Int J Miner Process 74:S395–S403
8. Sander S, Scubert G, Jäckel HG (2004) The fundamentals of the comminution of metals in shredders of the swing-hammer type. Int J Miner Process 74S:S385–S393
9. Ron A, Penev K (1995) Disassembly and recycling of electronic consumer products: an overview. Technovation 15(6):363–374
10. Shi F, Kojovic T, Esterle JS, David D (2003) An energy-based model for swing hammer mills. Int J Miner Process 71:147–166
11. Schuberta G, Bernotatb S (2004) Comminution of non-brittle materials. Int J Miner Process 74(10):S19–S30
12. Cui J, Forssberg E (2003) Mechanical recycling of waste electric and electronic equipment: a review. J Hazard Mater 99:243–263
13. Kasper AC (2011) Caracterização e reciclagem de materiais presentes em sucatas de telefones celulares. Universidade Federal do Rio Grande do Sul (UFRGS), Porto Alegre-RS, 120p (Dissertação)
14. Gupta A, Yan D (2006) Mineral processing design and operation: an introduction. Elsevier, Amsterdam
15. Luz AB, Sampaio JA, Salvador LMA (2004) Tratamento de Minérios. Rio de Janeiro. Cetem, 4° Edition
16. Zhang S, Forssberg E (1997) Mechanical separation-oriented characterization of electronic scrap. Resour Conserv Recycl 21:247–269
17. Goosey M, Kellner R (2002) A scoping study end-of-life printed circuit boards. Shipley Europe Limited, Coventry
18. Meili L, Bertuol DA (2011) Aplicação de Processamento Mecânico na Reciclagem de Resíduos Tecnológicos. Aplicações Tecnológicas em Sistemas Particulados, São Paulo, 157–183
19. Sampaio CH, Tavares LMM (2005) Beneficiamento Gravimétrico. Editora UFRGS
20. Wang B, Chu KW, Yu AB, Vince A (2014) Computational investigation of the mechanisms of the "breakaway" effect in a dense medium cyclone. Miner Eng 62:111–119
21. Rao LS, Bandopadhyay P (1992) Application of a Mozley mineral separator for treatment of coal washery rejects. Int J Miner Process 36:137–150
22. Mukherjee AK, Mishra BK (2006) An integral assessment of the role of critical process parameters on jigging. Int J Miner Process 81(3):187–200
23. Tsunekawa M, Kobayashi R, Hori K, Okada H, Abe N, Hiroyoshi N, Ito M (2012) Newly developed discharge device for jig separation of plastics to recover higher grade bottom layer product. Int J Miner Process 114–117:27–29

24. Carvalho MT, Agante E, Durão F (2007) Recovery of PET from packaging plastics mixtures by wet shaking table. Waste Manag 27(12):1747–1754
25. Eswaraiah C, Kavitha T, Vidyasagar S, Narayana SS (2008) Classification of metals and plastics from printed circuit boards (PCB) using air classifier. Chem Eng Process 47:565–576
26. Sivamohan R, Forssberg E (1985) Principles of tabling. Int J Miner Process 15:281–295
27. Zhang S, Forssberg E, Menad N, Bjorkman B (1998) Metals recycling from electronic scrap by air table separation-theory and application. In: EPD congress, pp 497–515
28. Rao SR (2006) Resources, recovery and recycling metallurgical wastes, vol 7. Elsevier, Amsterdam
29. Kasper AC, Berselli GBT, Freitas BD, Tenório JAS, Bernardes AM, Veit HM (2011) Printed wiring boards for mobile phones: characterization and recycling of copper. Waste Manag 31:2536–2545
30. Mihalk P (1979) Magnetic separation theory equipment and application. In: Word Mining: 105–106
31. Young HD, Freedman RA (2009) Fisica III: eletromagnetismo. Young & Freedman, 12° edn. Pearson Addison Wesley, São Paulo
32. Luz AB, Sampaio JA, França SCA (2010) Tratamento de Minérios - CETEM, 5º Edição Rio de Janeiro, 367–393
33. Svoboda JE, Fujita T (2003) Recent developments in magnetic methods of material separation. Miner Eng 16:785–792
34. Chen L (2011) Effect of magnetic field orientation on high gradient magnetic separation performance. Miner Eng 24:88–90
35. Franzreb M, Habich U (2002) Development of new permanent magnet-based separator for recycling of ferrous microparticles from aqueous effluents. In: Recycling and waste treatment in mineral and metal processing: technical and economic aspects, Lulea, Sweden, 16–20 June 2002
36. Zhao Y, Xi B, Li Y, Wang M, Zhu Z, Xia X, Zhang L, Wang L, Luan Z (2012) Removal of phosphate from wastewater by using open gradient superconducting magnetic separation as pretreatment for high gradient superconducting magnetic separation. Sep Purif Technol 86:255–261
37. Thurm S, Odenbach S (2002) Magnetic separation of ferrofluids. J Magn Magn Mater 252:247–249
38. Sierra C, Martínez J, Menéndez-Aguado JM, Afif E, Gallego JR (2013) High intensity magnetic separation for the clean-up of a site polluted by lead metallurgy. J Hazard Mater 248–249:194–201
39. Kurose K, Okuda T, Nishijima W, Okada M (2006) Heavy metals removal from automobile shredder residues (ASR). J Hazard Mater B137:1618–1623
40. Veit HM, Diehl TR, Salami AP, Rodrigues JS, Bernardes AM, Tenório JAS (2005) Utilization of magnetic and electrostatic separation in the recycling of printed circuit boards scrap. Waste Manag 25:67–74
41. Yoo JM, Jeong J, Yoo K, Lee JC, Kim W (2009) Enrichment of the metallic components from waste printed circuit boards by a mechanical separation process using a stamp Mill. Waste Manag 29:1132–1137
42. Tenório JAS, Espinosa DCR (2002) Recovery of Ni-based alloys from spent NiMH batteries. J Power Sources 108:70–73
43. Bertuol DA, Bernardes AM, Tenório JAS (2006) Spent NiMH batteries: characterization and metal recovery through mechanical processing. J Power Sources 160:1465–1470
44. Xua J, Thomas RH, Francis RW, Lum KR, Wang J, Liang B (2008) A review of processes and technologies for the recycling of lithium-ion secondary batteries. J Power Sources 177:512–527
45. Huang K, Li J, Xu Z (2011) Enhancement of the recycling of waste Ni–Cd and Ni–MH batteries by mechanical treatment. Waste Manag 31:1292–1299

46. Kelly EG, Spottiswood DJ (1989) The theory of electrostatic separations: a review, part I. Fundamentals. Miner Eng 2:33–46
47. Kelly EG, Spottiswood DJ (1989) The theory of electrostatic separations: a review, part II. Particle charging. Miner Eng 2:193–205
48. Li J, Lu H, Guo J, Xu Z, Zhou Y (2007) Recycle technology for recovering resources and products from waste printed circuit boards. Environ Sci Technol 41:1995–2000
49. Lungu M, Schlett Z (2001) Vertical drum eddy-current separator with permanent magnets. Int J Miner Process 63:207–216
50. Kelly EG, Spottiswood DJ (1989) The theory of electrostatic separations: a review, part III. The separation of particles. Miner Eng 2:337–349
51. Li J, Lu H, Xu Z, Zhou Y (2008) Critical rotational speed model of the rotating roll electrode in corona electrostatic separation for recycling waste printed circuit boards. J Hazard Mater 154:331–336
52. Hongzhou L, Li J, Guo J, Xu Z (2008) Movement behavior in electrostatic separation: recycling of metal materials from waste printed circuit board. J Mater Process Technol 197:101–108
53. Li J, Xu Z, Zhou Y (2007) Application of corona discharge and electrostatic force to separate metals and nonmetals from crushed particles of waste printed circuit boards. J Electrostat 65:233–238
54. Wu J, Qin Y, Zhou Q, Xu Z (2009) Impact of nonconductive powder on electrostatic separation for recycling crushed waste printed circuit board. J Hazard Mater 164:1352–1358
55. Wu J, Li J, Xu Z (2008) Electrostatic separation for multi-size granule of crushed printed circuit board waste using two-roll separator. J Hazard Mater 159:230–234
56. Jiang W, Jia L, Zhen-Ming X (2009) A new two-roll electrostatic separator for recycling of metals and nonmetals from waste printed circuit board. J Hazard Mater 161:257–262
57. Li J, Lu H, Liu S, Xu Z (2008) Optimizing the operating parameters of corona electrostatic separation for recycling waste scraped printed circuit boards by computer simulation of electric field. J Hazard Mater 153:269–275

Chapter 5
Leaching Processes

Camila Ottonelli Calgaro, Eduardo Hiromitsu Tanabe, Daniel Assumpção Bertuol, Flávia Paulucci Cianga Silvas, Denise Crocce Romano Espinosa and Jorge Alberto Soares Tenório

When solid materials are exposed to a liquid, some constituents will dissolve to a greater or lesser extent. The degree of dissolution of individual constituents by the contacting liquid leads to a leachate/percolate or extract composition that can be of interest for different purposes [1], such as the extraction of metals from solid waste through a suitable leachant. Leaching is a step in hydro- and electrometallurgical processes and has been also employed for the recovery of metals from waste of electrical and electronic equipment (WEEE). The process may be preceded by a mechanical pretreatment of the waste (see Chap. 6) and followed by purification and metal recovery from the solution (see Chaps. 8 and 9). This chapter will discuss the different leaching solutions and conditions for WEEE processing.

5.1 Acid and Alkaline Leaching

Camila Ottonelli Calgaro, Eduardo Hiromitsu Tanabe
Daniel Assumpção Bertuol

Laboratory of Environmental Process (LAPAM), Center of Technology, Universidade Federal de Santa Maria (UFSM), Santa Maria, RS, Brazil
e-mail: dbertuol@gmail.com

Leaching is the extraction of a soluble component from a solid by means of a solvent. The solvent and the solid are put into contact so the solute or solutes diffuse from the solid to the solvent, separating the components that constitute the original solid material [2]. The process can be used both for the production of a concentrated solution of a valuable solid, or it can be used to remove an insoluble solid

H.M. Veit and A. Moura Bernardes (eds.), *Electronic Waste*, Topics in Mining, Metallurgy and Materials Engineering, DOI 10.1007/978-3-319-15714-6_5

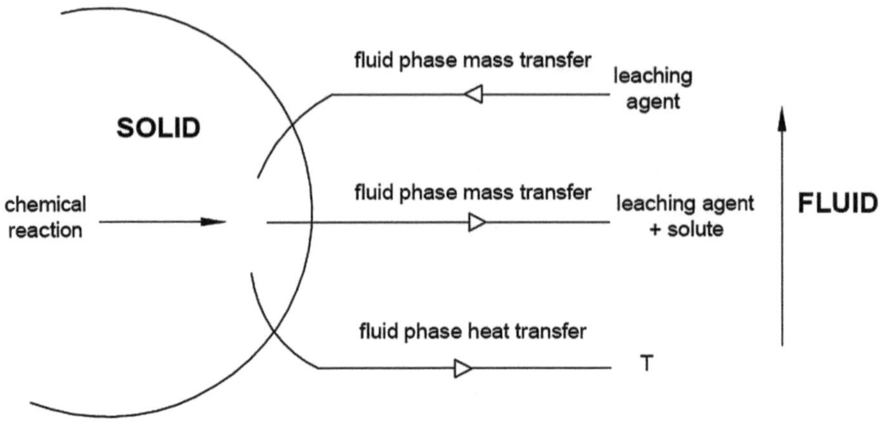

Fig. 5.1 Description of the mass transfer involved in the leaching process. Adapted from Sohn and Wadsworth [4]

that is contaminating a soluble material. The method to be used for extraction is determined by the proportion of the soluble component, its distribution throughout the solid, the nature of the solid and the particles size [3].

The mass transfer that occurs in the leaching process is described in Fig. 5.1. It involves the mass transfer from the leaching fluid into the solid, the mass transfer from the solute to the fluid, and from the fluid to the outer surface of the solid particle; the diffusion of reagents (leaching agents) and of the products (formed in the leaching reaction) within the pores of the solid, and the chemical reaction between the solute and the leaching fluid. There are also two processes that can significantly influence the mass transfer during leaching: one is heat transfer and the other is the change in structure during the solid reaction [4].

Considering that the rate of mass transfer from the surface of the solid to the liquid controls the leaching process, and ignoring the solid phase resistance [2], the rate of the solute being dissolved in the liquid bulk can be approximated using the thin film concept, as shown in Eq. 5.1 [3].

$$\frac{dM}{dt} = \frac{k_m A(C_s - C)}{\delta} \tag{5.1}$$

where:

A is the area of the solid-liquid interface,
δ is the effective thickness of the liquid film surrounding the particles,
C is the concentration of the solute in the bulk solution at time t,
C_s is the concentration of the saturated solution in contact with the particles,
M is the mass of solute transferred at time t, and
k_m is the diffusion coefficient.

Considering the process as a batch process at a constant volume of solution (V), Eq. 5.2 can be written as:

$$dM = VdC \tag{5.2}$$

Combining Eqs. 5.1 and 5.2, we get:

$$\frac{dC}{dt} = \frac{k_m A(C_s - C)}{\delta V} \tag{5.3}$$

Through the integration of Eq. 5.3 from C_0 (initial concentration of solute in the liquid) to C (concentration reached at time t) Eq. 5.4 is obtained. Equation 5.4 allows us to infer that the concentration of the solution approaches the saturated condition exponentially. In most cases the interfacial area tends to increase during extraction and when the soluble material is a large proportion of the solid, the complete disintegration of the particles can occur [3].

$$\frac{C_s - C_0}{C_s - C} = e^{\frac{k_m A}{\delta V} t} \tag{5.4}$$

The selection of the equipment to carry out the leaching process must consider the factors that limit the rate of extraction. When the diffusion of the solute through the porous structure of the solid limits the process, the solid particles must be reduced in size to facilitate the diffusion of the solid. However, when the diffusion of solute from particle surface to solvent medium limits the leaching, a high degree of stirring of the fluid is required. There are four main factors that influence the process:

- Particle size: the smaller the particle size, the larger the interfacial area between the solid and the solvent and, consequently, the higher the rate of mass transfer and the smaller the distance that solute must diffuse into the solid. However, a minimum particle size should be respected, avoiding very small particles from depositing in the interstices of larger particles and, as suchm, hindering the solvent flow;
- Solvent: the solvents should be selective to the desired components and have low viscosity, favoring their circulation;
- Temperature: in most cases the increase in temperature increases the solubility of the material being extracted;
- Stirring of the fluid: This increases the mass transfer from the surface of the particles, thus favoring the extraction [3].

Leaching is often employed in hydrometallurgical processes. Since the metals generally occur in mixtures with undesirable constituents, leaching is used to get them as soluble salts. For example, copper salts are leached from ores containing other minerals with sulfuric acid or ammoniacal solutions. Since the solvent used in the leaching depends on the desired solute, there must be an affinity between them. When the solute is enclosed by an insoluble matrix, it must diffuse inside the particle, dissolve the components of interest and spread out again to the solid matrix. This occurs in many hydrometallurgical processes where the metal salts are leached from ores and grinding operations are required to increase the rate of leaching and facilitate contact of the solvent with the solute [2].

Leaching is a step in hydrometallurgical processes employed for ores and that has also been employed in the recovery of metals from WEEE [5, 6].

The development and use of the leaching process as a step in the recycling of WEEE was inspired by, or even originated from the extraction of metals from ores with acid or alkaline leaching, which is already well established on an industrial scale. Before its development for recycling, it is also important therefore to understand the leaching procedure employed in the extraction of metals from ores.

Heap leaching is a method to extract metals from ores that is used in the extraction of gold, silver and copper. The processed ore is piled on waterproof pads. Leaching solutions are applied on the surface of the heap ore by sprinkler or drip irrigation, leaching the metals contained in the ore. Sulfuric acid and cyanide are typically used as leaching agents.

This leaching process, which involves the percolation of leaching solutions through ore heaps, collects a solution enriched with metals by gravity in lined basins, sumps, and ponds for further processing and recovery of metals. The lixiviant solution is recovered and recycled in the process in a closed system [7].

The leaching parameters for the development and use of leach heaps are guided by mineralogy, which considers aspects such as dissolved oxygen, particle size, metal distribution into particles (on the surface or inside) and porosity of the particles. In order to optimize the leaching process, it is essential to have quantitative information about the effect of the solid size and the effect of the percolation leaching within the solid. Smaller particles usually favor quicker leaching kinetics due to a greater surface area.

These characteristics directly influence the required degree of grinding for efficient metal release, the processing techniques required and the nature and amount of leaching solution to be used. Since the amount of leaching agent is critical to the process and since it represents one of the largest cost items, characteristics of the ore are relevant to the extraction process [8].

The metal recovery from WEEE through the leaching process typically requires oxidative leaching [5, 6].

Acid leaching is often used as the first leaching stage for the extraction of base metals such as copper. The leaching efficiency of different oxidants and mineral acids, such as hydrochloric acid, sulfuric acid, nitric acid/hydrogen peroxide, perchloric acid and sodium hypochlorite, has been studied. Hydrogen peroxide is a strong oxidizer, so it is combined with acids to improve the leaching performance [6].

However, for the recovery of precious metals from waste, such as gold, silver and palladium, the following chemicals compounds are employed:

- Cyanide: already used in different plants as leaching agent for gold recovery from ore. Effectively dissolves the precious metals in basic pH of around 10–10.5. The mechanism of metal dissolution in cyanide is essentially electrochemical, as shown in Eqs. 5.5 and 5.6 [5].

$$4Au + 8CN^- \rightarrow 4Au(CN)_2^- + 4e \tag{5.5}$$

$$O_2 + 2H_2O + 4e \rightarrow 4OH^- \quad (5.6)$$

Cyanide is a very efficient leaching agent mainly for gold, but it represents a potential risk to the environment because of its high toxicity, [6].

- Halogens: with the exception of fluorine and astatine, all the halogens have been tested or used for the extraction of gold. However, only chloride is employed in industrial scale. The extraction rate using chloride as leachant is favored at low pH, with increased temperature and large surface areas. The leaching traditionally used to extract the precious metals is the aqua regia (formed by mixing concentrated nitric acid and hydrochloric acid in a volume ratio of 1:3). Equations 5.7 and 5.8 describe its use in gold extraction [5]. However, the application of chlorine in gold leaching is hampered by its corrosive power, which requires adjustments to the equipment, and by its high toxicity in the gaseous state [9].

$$2HNO_3 + 6HCl \rightarrow 2NO + 4H_2O + 3Cl_2 \quad (5.7)$$

$$2Au + 11HCl + 3HNO_3 \rightarrow 2HAuCl_4 + 3NOCl + 6H_2O \quad (5.8)$$

Park and Fray [10] investigated the influence of the solid-liquid ratio of aqua regia on the recovery of Au from printed circuit boards (PCB), confirming the high solubility of gold in aqua regia. The authors reported a minimum gold recovery rate of 80 %, reaching 100 % in the highest tested ratio, for leaching times of 3 h at a temperature 20 °C on three solid-liquid ratios, 1 g:10 mL, 1 g:20 mL and 1 g:40 mL.

- Thiourea, also known as sulfur urea, is an organic complexing agent, which under acidic conditions and in the presence of an oxidant, typically Fe^{3+}, forms a soluble cationic complex with the gold [11]. This is a rapid extraction reaction (described in Eq. 5.9) and the rate of gold recovery can reach 99 % [5]. In the recycling of printed circuit boards, for example, thiourea shows good selectivity and efficiency for the recovery of gold and silver. However, when compared with other leaching agents, thiourea has low stability [11]. Its industrial use is still limited, mainly because it is more expensive and also because it uses a higher amount of reactants than cyanide. As such, more studies are still required [5].

$$Au + 2CS(NH_2)_2 \rightarrow Au(CS(NH_2)_2)_2^+ + e \quad (5.9)$$

Birloaga et al. [12] extracted 82.1 % of Au from CPU waste with 20 g/L of Thiourea, 6 g/L of Fe^{3+}, 10 g/L of H_2SO_4, with agitation at 600 rpm and a particle size of 0.1 mm for 3.5 h, confirming the efficiency of the leaching. The influence of the particle size was established, since for particles bigger than 3 mm, the extraction was only 18.2 %.

- Thiosulfate: two types of thiosulfate are used in gold leaching; sodium thiosulfate and ammonium thiosulfate. Gold and thiosulfate form a stable complex in the presence of oxygen [11]. The dissolution of gold in a ammonium thiosulfate solution is characterized by an electrochemical reaction catalyzed by the presence of copper ions (as described in Eqs. 5.10 and 5.11)

[5]. Thiosulfate is stable in mildly alkaline media. Its use has advantages, such as high selectivity for precious metals, non-toxicity and non-corrosiveness [11]. One disadvantage, however, is that large quantities are needed for leaching [5].

$$Au + 5S_2O_3^{-2} + Cu(NH_3)_4^{2+} \rightarrow Au(S_2O_3)_2^{-3} + 4NH_3 + Cu(S_2O_3)_3^{-5} \quad (5.10)$$

$$2Cu(S_2O_3)_3^{-5} + 8NH_3 + \frac{1}{2}O_2 + H_2O \rightarrow 2Cu(NH_3)_4^{2+} + 2OH^- + 6S_2O_3^{-2} \quad (5.11)$$

Different leaching efficiency results are presented in the literature. Ha et al. [13], using thiosulfate 0.12 M, ammonia 0.2 M and copper 20 mM, recovered 98 % of the gold contained in the keyboards of mobile by leaching within 2 h. Maintaining the same conditions, the authors extracted 90 % of the gold present in the printed circuit boards of cell phones in a time of 10 h. On the other hand, Petter et al. [14], studying the leaching of gold and silver from the PCBs of mobile phones with sodium and ammonium thiosulfate, extracted only 15 % of the gold present in the PCBs. The authors concluded that for the recovery of metals present in the PCBs, more than one leaching solution would be needed.

PCBs present in WEEE have a significant amount of valuable metals in their composition, especially copper and precious metals. The presence of these metals helps to encourage recycling and the development of more efficient recycling methods for this kind of waste [15–17].

Table 5.1 presents different leaching solutions used in the recycling process of PCBs. These compounds interact in different ways with the metals, enabling a greater or smaller extraction of a certain metal, depending on the leachant and the conditions used in the process [6].

Another leaching application for the recovery of metals that has been studied is the recycling of secondary lithium ion batteries, for which demand and disposal is increasing because of their fast technological development. The cathode of Li-Ion batteries has lithium and cobalt in its composition, which are the main motivators for their recycling [23]. The leaching of the cathodic material ($LiCoO_2$) is accomplished through the use of the inorganic acids H_2SO_4, HCl and HNO_3, with the addition of H_2O_2 [24]. Studies on the use of these leaching solutions consider the effect of the following parameters: concentration, temperature, leaching time and solid-liquid ratio [25].

Previous research has shown that when H_2SO_4 is employed as leaching agent of lithium ion batteries, the increase in the temperature and in the acid concentration, in addition to the reduction of the solid/liquid (S/L) ratio to certain values, and particularly the addition of H_2O_2, increases the extraction rate of lithium and cobalt [26, 27]. For example, with sulfuric acid 2 M at 75 °C, a S/L ratio of 100 g/L, H_2O_2 5 % (v/v) and a leaching time of 60 min, it was possible to recover 99.1 % of lithium and 70 % of cobalt [28]. In another study where the H_2O_2 percentage added was 15 % (v/v) and the S/L ratio was reduced to 50 g/L, with all other conditions equal to the previous ones, the recovery of cobalt and lithium

Table 5.1 Leaching solutions employed in the recovery of different metals from PCBs wastes

Scrap type	Leaching reagent	Conditions	Recycled metals	Reference
PCB	1st leaching step: H_2SO_4 (2M) + H_2O_2	25 °C, 3 h	Cu (85.76 %), Ag (0.86 %)	[18]
	2nd leaching step: H_2SO_4 (2M) + H_2O_2	25 °C, 3 h	Cu (13.99 %), Ag (11.30 %)	
	3rd leaching step: thiourea (20 g/L) + ferric iron (6 g/L) + H_2SO_4 (10 g/L)	25 °C, 3 h	Au (84.31 %), Ag (71.36 %)	
PCB	Aqua regia	Density 1 g:20 mL, 20 °C, 3 h	Ag (2 %), Pd (7 %), Au (97 %)	[10]
PCB	H_2SO_4 (2.18N)	Density 1 g:10 mL, 60 °C, 2 h	Sn (2.7 %), Cu (<0.01 %)	[19]
	H_2SO_4 (2.18N) + HCl (3N)		Sn (59.3 %), Cu (8.9 %)	
	HCl (3N)		Sn (89.1 %), Cu (33.2 %)	
	HCl (3N) + HNO_3 (1N)		Sn (98.1 %), Cu (93.2 %)	
PCB	H_2SO_4 (2M) + H_2O_2 (0.2M)	85 °C, 12 h	Cu, Fe, Zn, Ni, Al (>95 %)	[20]
	$(NH_4)_2S_2O_3$ (0.2M) + $CuSO_4$ (0.02M) + NH_4OH (0.4M)	40 °C, 48 h (Au), 24 h (Ag)	Au (>95 %), Ag (100 %)	
PCB	H_2SO_4 (0.5M) + Cu^{2+}(4 g/L) + Cl^-(46.6 g/L)	80 °C, 0.5 h, density 1 % w/v	Cu (92 %)	[21]
PCB (soldering material on the outer layer epoxy resin)	1st leaching step: HNO_3 (0.2M)	90 °C, 45 min, density 1 g/100 mL	Pb (99.99 %)	[22]
	2nd leaching step: HCl (3.5M)	90 °C, 2 h, density 1 g/20 mL	Sn (98.74 %)	

reached values close to 100 %, demonstrating the relevance of these variables for the process optimization of the process [25].

Leaching is also applied in the recycling of glass from cathode ray tubes (CRT), which contains high amounts of heavy metals and cannot be discarded without prior treatment, especially due to the presence of a significant amount of Pb [29]. Leaching is an important step in the recycling CRTs, and it is favored by higher temperatures, stirring, reduced particle size (large surface area), a high S/L ratio and a high solvent concentration [30]. Using such leachants as nitric acid 0.1 M or acetic acid 0.5 M, with a S/L ratio of 0.5 %, it is possible to extract more than 95 % of Pb [31].

5.2 Leaching Using Supercritical Fluids

Camila Ottonelli Calgaro, Eduardo Hiromitsu Tanabe, Daniel Assumpção Bertuol

Laboratory of Environmental Process (LAPAM), Center of Technology, Universidade Federal de Santa Maria (UFSM), Santa Maria, RS, Brazil
e-mail: dbertuol@gmail.com

The use of supercritical fluids in the recovery of metals from solid and liquid matrices is presented today as a promising alternative to the techniques that are conventionally employed because of their favorable solvent properties compared to other solvents [32].

Supercritical fluids are substances that are submitted to pressures and temperatures that exceed their critical points (Fig. 5.2).

Supercritical fluids have higher solvent power for non-polar or slightly polar components and are easily removed from the solutes by simple expansion at room pressure. In the supercritical state, the density of the fluid approaches the density of the liquids, while the viscosity is close to that of normal gases, and the diffusivity of the fluid is about two orders of magnitude greater than the typical liquid diffusivity (Table 5.2) [33]. The low viscosity and high diffusivity of the supercritical fluid cause significant density gradients at the solute/solvent interface, resulting in higher mass transfer rates than with conventional solvents [34].

The basic principles to be considered for the use of a supercritical fluid as an extractor medium are the solubility and the equilibrium phase of the substances in the compressed gas. First, the components that have to be extracted must be

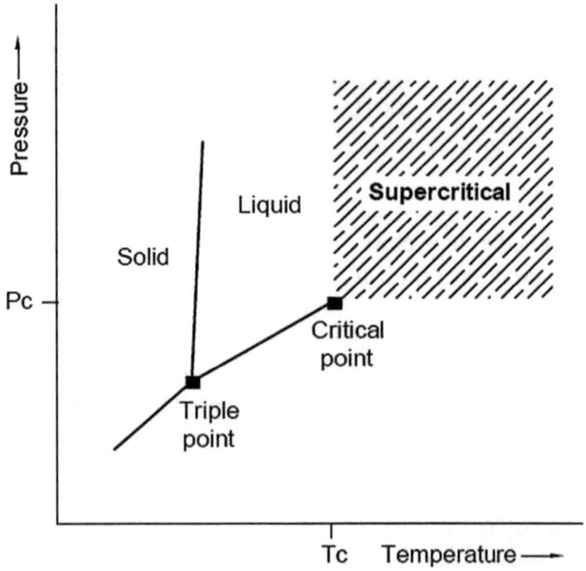

Fig. 5.2 Supercritical state of a pure component. Adapted from Brunner [33]

Table 5.2 Characteristic values of different states: gas, liquid and supercritical fluid

State of fluid	Density (g/cm^3)	Diffusivity (cm^2/s)	Viscosity (g/cm s)
Gas			
p = 1 atm, T = 15–30 °C	$(0.6–2) \times 10^{-3}$	0.1–0.4	$(0.6–2) \times 10^{-4}$
Liquid			
p = 1 atm, T = 15–30 °C	0.6–1.6	$(0.2–2) \times 10^{-5}$	$(0.2–3) \times 10^{-2}$
Supercritical fluid			
$p = p_c$; $T \approx T_c$	0.2–0.5	0.7×10^{-3}	$(1–3) \times 10^{-4}$
$p = 4p_c$; $T \approx T_c$	0.4–0.9	0.2×10^{-3}	$(3–9) \times 10^{-4}$

Adapted from Brunner [33]

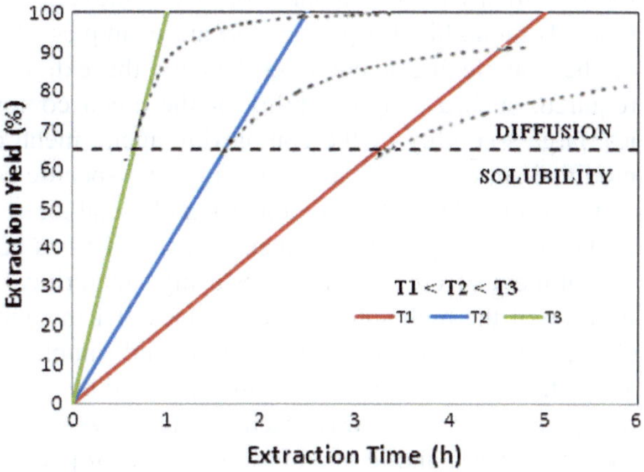

Fig. 5.3 Typical behavior of the extraction process as a function of time. Adapted from Laurent et al. [35]

soluble in the supercritical fluid at moderate temperatures and pressures. The solubility and phase-equilibria can be modified by altering pressure and temperature. At the beginning of the extraction process, the efficiency is limited by the solubility of the compounds in the available fluid quantity (Fig. 5.3).

High solubility and, consequently, lower extraction times can be achieved by increasing the extraction pressure, since the increase of the fluid density causes an increase in the solvent power. The same effect is normally achieved by increasing the extraction temperature, since the effect of increasing the vapor pressure of the components being dissolved overcomes the effect of reducing the density of the fluid at high temperatures. The second phase of extraction is controlled by diffusion, which is slow. So in order to make extraction with supercritical fluids economically feasible, the goal should be to attain the required extraction rate during the solubility phase [35].

When a pure supercritical solvent has low solubility for the compounds of interest, it is possible to affect the polarity of the fluid by adding small amounts of modifiers in the gas extractor, enabling an increase of solubility of polar substances [35].

CO_2 is one of the most used supercritical solvents because it has moderate critical conditions (30.9 °C and 73.8 bar), is inexpensive, environmentally acceptable, considered safe by the U.S. Food and Drug Administration (FDA) and European Food Safety Authority (EFSA), has high diffusivity and is gaseous at room temperature and pressure (facilitating separation and recovery of the solute) [32]. Such characteristics as the high diffusivity and low viscosity of supercritical CO_2, allow for direct extraction of solutes from the solid matrices [36]. However, its main disadvantage is the low polarity, which often makes the addition small quantities of modifiers necessary, such as methanol, which allows the extraction of polar compounds [32].

When CO_2 is the supercritical fluid extractor for the extraction of metals, complexing or chelating agents are used for the complexation of metal ions, favoring their separation [32]. Chelating or complexing converts the metal species into neutral complexes that are soluble in supercritical CO_2 [37]. Cyanex 302, sodium diethylthiocarbamate, bis(2-ethylhexyl)phosphoric acid and bis(2-ethylhexyl) monothiophosphoric acid are examples of such agents [32]. The choice of the complexing agent is critical for the extraction. Some characteristics are required, such as: high stability of the metal complex, fast complexation kinetics, large solubility of the complexing agent, metal complexes formed in pure or modified supercritical carbon dioxide, and specific complexing, to allow selective extraction for the metal ion or a group of metal ions [38].

The use of the chelating agent Cyanex 301 has been investigated in the extraction of metals from solid matrices using supercritical CO_2 modified with methanol because of the high solubility of Cyanex 301 in supercritical CO_2 [38, 39]. Using Cyanex 301 in a synthetic mixture of solid samples of Ag, Au, Bi, In, Mn and Pd, a significant extraction of these metals could be achieved. For example, in the case of Ag, a rate of 92 % was reached using CO_2 modified with methanol, at a pressure of 300 atm and temperature of 318 K [38]. The extraction efficiency depends particularly on the characteristics of the complexing or chelating agent, the temperature and pressure conditions, the complexes formed with the metals, pH and the characteristics of the matrix [40].

In addition to CO_2, water has been employed in supercritical conditions ($T > 374$ °C, $P > 22.1$ MPa) for the recycling of waste from electrical and electronic equipments. More specifically, for the pre-treatment of printed circuit boards (PCB). The supercritical water is used to remove the organic fraction present in PCBs, allowing the separation of the solid phase rich in metals. There are two types of processes that use supercritical water to treat discarded PCBs: Supercritical water oxidation (SCWO) in the presence of oxygen and supercritical water depolymerization (SCWD) under a reducing atmosphere. During the pretreatment with supercritical water, three phases are formed: solid, liquid and gas, and the metals remain in the solid or in the aqueous phase [41].

When the SCWO process is used as a pretreatment in the recycling of PCBs, Cu and Pb are oxidized to CuO, Cu_2O and PbO_2, facilitating their recovery. The optimal conditions are 440 °C, 30 MPa (above the critical point of water) for 60 min [42]. Pretreatment with SCWO has a high efficiency for the recovery of Cu and Pb, and the efficiency increases significantly with increasing temperature in the pretreatment, since the pretreatment with SCWD positively influences the recovery of Sn, Cd, Zn and Mn [41].

The application of methanol in supercritical conditions in the recycling of PCBs has also been studied, because methanol has milder supercritical conditions (T > 240 °C, P > 8.09 MPa) than water, and because the separation between products and solvent is easier as a result of the lower boiling point of methanol when compared to water. Supercritical methanol is used to recover both polymers and metals from PCB waste. Some parameters have a significant effect on this extraction: temperature and fluid pressure, extraction time and S/L ratio are the main ones. At temperatures ranging from 300 to 420 °C, and with extraction times between 30 and 120 min, a S/L ratio from 1:10 to 1:30 and pressures between 9 and 21 MPa, it is possible to recover organic compounds, mainly phenol, and especially metals like Cu, Fe, Sn, Pb and Zn, and to a smaller degree, Ag and Au [43]. The supercritical fluid technology used for the recovery of metals and for the recycling of waste from electrical and electronic equipments is therefore promising, but more studies are still needed for further analysis and comparison with conventional hydrometallurgical methods.

5.3 Bioleaching

Flávia Paulucci Cianga Silvas, Denise Crocce Romano Espinosa, Jorge Alberto Soares Tenório

Department of Chemical Engineering, Polytechnic School, University of São Paulo (USP), São Paulo, Brazil
e-mail: flavia.silvas@gmail.com

The active participation of microorganisms in geochemical processes has made the development of technological processes possible that fall under the generic name of biohydrometallurgy. These processes are based on chemical and/or physical interactions between microorganisms, their by-products and the substrate. One of these processes is bioleaching, where metals from inorganic sources (like minerals, for example) are solubilized by the action of microorganisms [44].

Notwithstanding the recent application of bioleaching in the treatment of electrical and electronic waste, this method has been in use since 1700 when heap leaching was commercially established in copper mines. But the participation of microorganisms in the process was unknown [45].

At the end of the 19th century, studies showed that bacteria could possibly survive by obtaining their energy from the oxidation of inorganic substrates, fixing atmospheric carbon dioxide. During the 20th century more in-depth research was carried out in this field, and the effective participation of microorganisms in the leaching of ore was established [46].

The bioleaching of ore is currently used on an industrial scale, mainly to win copper and gold [45]. It can also be used, however, to recover other metals from low concentration ore or waste, such as uranium, cobalt, molybdenum, nickel and lead [47].

In the 1970s, 18 % of the production of copper in the United States came from bioleaching processes. Nowadays, it is estimated that 15 % of the world's copper production from comminuted waste and ore comes from bioleaching [48].

Recently, the leaching of secondary sources of metals through the action of microorganisms has been researched extensively, and some promising results are starting to appear. Among the waste used as raw material are sludge from galvanization plants [49], ashes [50–52], sediments [53], batteries [54–58], electrical and electronic waste [59–65], catalysts [66, 67, 68], mixed metallic waste ("fine shredder residues") [69], red mud [70, 71], waste from copper smelting [60, 72] and soil [73, 74].

Bioleaching is seen as one of the most promising technologies for the treatment and recovery of materials from WEEE [75], since it is considered to be a low-cost process, both in terms of its installation and operation, with low energy consumption and the use of non-toxic reagents [76].

Some factors must be taken into consideration for a successful bioleaching process, such as the presence of toxic substances, particle size and surface area of the material to be bioleached, prior adaptation of the microorganisms, temperature, pH, oxygen reduction potential (Eh), available O_2 and CO_2, amount of nutrients and the form in which the metals are present (oxides, sulfides, metallic form). One must therefore know the properties and characteristics of the waste to be able to determine the best microorganisms and conditions to be used [44, 77, 78].

Acidithiobacillus ferrooxidans were the first microbial species discovered that were capable of oxidizing and solubilizing mineral sulfides and they are still one of the main microorganisms used in the conventional bioleaching of sulfide minerals. Other acidophilus bacteria of the *Acidithiobacillus* genus, mainly the species *thiooxidans*, but also the *Leptospirillum ferrooxidans* bacteria, are also commonly used in traditional processes. Other microorganisms, however, can also be used, such as: *Sulfolobos acidocaldarius*, *Metallogenium* sp., *Gallionella* sp., *Leptothrix* sp., *Crenothriz* sp. and *Sulfobacillus thermosulfidooxidans.* Different species can also be used at the same time through a cooperative interaction that promotes mutual benefits [59, 64, 79].

In the traditional bioleaching of sulfide ore, the oxidation reaction is catalyzed by the action of microorganisms (MOOS) that are capable of oxidizing Fe^{2+} and/or sulfonated compounds (Eq. 5.12, general process reaction) [80–83].

$$MS_{(s)} + 2O_{2(g)} \xrightarrow{MOOS} MSO_{4(aq)} \tag{5.12}$$

The oxidation of a metallic sulfide by bacterial action can occur by way of two mechanisms: direct and indirect [80, 81].

In direct leaching, there is an interaction between the bacteria (enzymatic system) and the surface of the mineral sulfide without the formation of ferrous sulfate. Initially, the adherence of the microorganisms to the substrate occurs by van der Waals-type interaction forces. The contact of the MOOS with the surface of the substrate intensifies the formation of extracellular compounds produced by the bacteria, the exopolysaccharides (EPS). The adhered bacteria start reproducing and form micro-colonies, which in turn cover the material like a multi-layered film, known as biofilm [80].

The reactions that occur in the direct mechanism are represented by Eqs. 5.13 and 5.14.

$$M_2S_{(s)} + \frac{1}{2}O_{2(g)} + 2H^+_{(aq)} \xrightarrow{MOOS} M^{2+}_{(aq)} + S^0_{(s)} + H_2O \quad (5.13)$$

$$S^0_{(s)} + \frac{3}{2}O_{2(g)} + H_2O \xrightarrow{MOOS} H_2SO_{4(aq)} \quad (5.14)$$

In indirect leaching, the oxidation of the mineral sulfide is caused by the action of the Fe^{3+} ion, as shown in Eq. 5.15.

$$MS_{(s)} + 2Fe^{3+}_{(aq)} \rightarrow M^{2+}_{(aq)} + S^0_{(s)} + 2Fe^{2+}_{(aq)} \quad (5.15)$$

The Fe^{2+} ion is reoxidized to Fe^{3+} by the action of the bacteria (Eq. 5.16) and the regenerated Fe^{3+} ion feeds back into Eq. 5.15.

$$2Fe^{2+}_{(aq)} + \frac{1}{2}O_{2(g)} + 2H^+_{(aq)} \xrightarrow{MOOS} 2Fe^{3+}_{(aq)} + H_2O \quad (5.16)$$

Direct and indirect mechanisms can occur simultaneously. So the acid formed in the oxidation reaction of the elementary sulfur by the direct mechanism (Eq. 5.14) helps with the solubilization of the metallic sulfide by way of the indirect mechanism, shown in Eq. 5.17.

$$MS_{(s)} + 2H^+_{(aq)} \rightarrow M^{2+}_{(aq)} + H_2S_{(g)} \quad (5.17)$$

However, it is currently believed that during the bioleaching process, Fe^{3+} and/or protons are the only agents responsible for dissolving the metal sulfide, in a indirect mechanism. The bacteria are just responsible for regenerating the Fe^{3+} and/or protons and make the connection between the mineral and the bacterial cell to increase the degradation process. The oxidation reaction of the metal sulfides will also depend on the characteristics of their electronic structure wherein them can be divided into two major groups [80, 81]:

1. Metallic sulfides that are capable of transferring electrons directly from the metal of which they are composed to the Fe^{3+} ion and, after a series of oxidation reactions and loss of electrons, of breaking the chemical bond between the metal and the sulfur and of producing thiosulfate. This group includes, for example, FeS_2 (pyrite), MoS_2 (molybdenite) and WS_2 (tungstenite). The reactions that occur in this oxidative dissolution process of the pyrite are described in Eq. 5.18 (chemical reaction, formation of thiosulfate), and Eq. 5.19 (biological reaction, regeneration of thiosulfate and/or Fe^{3+}) and in Eq. 5.20 (alternative oxidation of thiosulfate)

$$FeS_{2(s)} + 6Fe^{3+}_{(aq)} + 3H_2O \rightarrow S_2O^{2-}_{3(aq)} + 6H^+_{(aq)} + 7Fe^{2+}_{(aq)} \quad (5.18)$$

$$S_2O^{2-}_{3(aq)} + 8Fe^{3+}_{(aq)} + 5H_2O \xrightarrow{MOOS} 2SO^{2-}_{4(aq)} + 8Fe^{2+}_{(aq)} + 10H^{+}_{(aq)} \quad (5.19)$$

$$2Fe^{3+} + 2S_2O_3^{2-} \leftrightarrow 2FeS_2O_3^{+} \rightarrow 2Fe^{2+} + S_4O_6^{2-} \quad (5.20)$$

2. Metallic sulfides that are partially soluble in acids provided that the electrons of the valence layer can be "stolen" both by the Fe^{3+} and by the protons, breaking the bond between the metal and the sulfur. In the reactions that occur in this process the metal is fully/partially solubilized in an acidic environment and oxidation may also occur by the action of the Fe^{3+} ion. Polysulfide is formed in the reaction that may be solubilized by the action of iron or sulfur-oxidizing microorganisms (Eqs. 5.21–5.23, respectively).

$$nMS_{(S)} + 2(n-1)Fe^{3+}_{(aq)} \rightarrow nM^{2+}_{(aq)} + S^{2-}_{n(aq)} + 2(n-1)Fe^{2+}_{(aq)} \quad (n \geq 2) \quad (5.21)$$

$$\frac{1}{2}H_2S_n + Fe^{3+}_{(aq)} \xrightarrow{MOOS} \left(\frac{1}{16}n\right)S_8 + Fe^{2+}_{(aq)} + H^{+}_{(aq)} \quad (5.22)$$

$$\frac{1}{8}S_{8(aq)} + \frac{1}{2}O_{2(g)} + H_2O \xrightarrow{MOOS} SO^{2-}_{4(aq)} + 2H^{+}_{(aq)} \quad (5.23)$$

The bioleaching process of electrical and electronic waste is similar to that of conventional sulfides, where the Fe^{2+} ion is oxidized by the action of acidophilus bacteria (for example: *Acidithiobacillus ferrooxidans* and *Leptospirillum ferrooxidans*) for regeneration of the Fe^{3+} ion, in accordance with the previously described Eq. 5.16 [76]. However, in electrical and electronic waste there are no metallic sulfides, so the microbial action is directed towards oxidizing and regenerating the Fe^{3+} (Eq. 5.24).

$$2Fe^{2+}_{(aq)} + \frac{1}{2}O_{2(g)} + 2H^{+}_{(aq)} \xrightarrow{MOOS} 2Fe^{3+}_{(aq)} + H_2O \quad (5.24)$$

The Fe^{3+} ion is a strong oxidizing agent (0.77 V) capable of oxidizing and solubilizing the metals in the waste (Eq. 5.25).

$$M^0 + 2Fe^{3+} \rightarrow 2Fe^{2+} + M^{2+} \quad (M = bivalentmetal) \quad (5.25)$$

So after dissolving the metal through oxidation, the Fe^{3+} ion is reduced to Fe^{2+}, which in turn is used as the source of energy for the bacteria and for regenerating the Fe^{3+}, thus feeding the Fe^{2+}—MOOS—Fe^{3+} cycle.

Generally speaking, electrical and electronic waste is alkaline in character, which helps to increase the pH and consequently facilitates the precipitation of iron as $Fe(OH)_3$ (Eq. 5.26). The presence of ferrous hydroxide, combined with the soluble NH_4^+ ion, may result in the formation of an ammonia-jarosite precipitate (Eq. 5.27) [84]. However, the formation of jarosite may also occur in the presence of other ions, such as Na^+, K^+.

$$Fe^{3+} + 3H_2O \rightarrow Fe(OH)_3 + 3H^+ \quad (5.26)$$

$$Fe(OH)_3 + \frac{4}{3}3SO_4^{2-} + Fe^{3+} + H_2O + \frac{2}{3}NH_4^+ \rightarrow \frac{2}{3}(NH_4)Fe_3(SO_4)(OH)_6 + H^+ \quad (5.27)$$

It is known that some microorganisms are capable of producing and excreting metabolic by-products, like polysaccharides, amino acids and proteins that help to dissolve the metals [85]. Among them are soluble organic acids with a low molecular weight, which in addition to acidifying the environment also act as chelating agents, assisting in the solubilization of the metals [75, 85].

The reactions involved in this metal dissolving process include the cathodic reduction reactions of hydrogen protons (Eq. 5.28), anodic oxidation of the metal (Eq. 5.29) and the formation of stable complexes between the radical of the organic acid and the metal (Eqs. 5.30–5.32) [75].

$$2H_3O^+ + 2e^- \rightarrow H_2 + H_2O \quad (5.28)$$

$$M \rightarrow M^{2+} + 2e^- \quad (M = bivalent\ metal) \quad (5.29)$$

$$M^{2+} + H_3R' = MR^- + 3H^+ \quad (5.30)$$

$$M^{2+} + H_2R' = MR + 2H^+ \quad (5.31)$$

$$M^{2+} + HR' = MR^+ + H^+ \quad (5.32)$$

(*R' = Radical with valency* 3^-. *For example, a citrate molecule deriving from citric acid*)

Unlike the conventional bioleaching process, in which autotrophic bacteria are the main microorganisms acting in dissolving the metals by organic acids, the main role is played by heterotrophic microorganisms, such as fungi and some bacteria. They are capable of metabolizing different acids, such as malic, lactic, oxalic, citric, fumaric, ketoglutaric, succinic and pyruvic acid. Fungi of the *Aspergillus* and *Penicillium genii*, especially the *A. niger* and *P. simplicissimum* species, are some of the most commonly used. Among the bacteria, the main genii are the *Bacillus* and *Pseudomonas* [86].

Recent research has shown that the action of acids in the bioleaching process is of influence on the solubilization rate of the metal [75, 87]. According to Saidan et al. [75], in a pH of 2, bio-metabolized organic acids lead to a better dissolution of the copper coming from printed circuit boards than sulfuric acid, which is the result of bacterial action. In a pH lower than 1.5, sulfuric acid gives better results.

Despite the growing number of studies involving the application of bioleaching for treating and recovering metals from secondary sources, its use in the real world is still limited. Among the types of waste researched are batteries, printed circuit boards and electrical and electronic waste in general. Table 5.3 contains a summary of some of the works, and of the technique, waste and microorganisms used.

Table 5.3 Studies in the bioleaching of electrical and electronic waste

Waste	MOOS	Technique	Reference
Lithium ion batteries	*Acidithiobacillus thiooxidans* and *Acidithiobacillus ferrooxidans*	Agitated flasks	[58]
	Sulfur-oxidizing and iron-oxidizing *Acidithiobacillus*		[56]
Zn–Mn batteries	*Alicyclobacillus* sp. and *Sulfobacillus* sp.		[57]
Ni–Cd batteries	*Acidithiobacillus thiooxidans*		[88]
	Acidithiobacillus sp.	Reactor	[54]
	Acidithiobacillus sp.	Continuous reactor	[55]
PCB	*Acidithiobacillus* sp., *Gallionella* sp. and *Leptospirillum* sp.	Agitated flasks	[62]
	Sulfobacilllus thermosulfidooxidan, *Thermoplasma acidophilum*, and *Sulfobacilllus acidophilus*		[63]
	Sulfobacilllus thermosulfidooxidans and *Thermoplasma acidophilum*	Columns	[65]
	Acidophilus bacteria	Agitated flasks	[59]
	Acidithiobacillus ferrooxidans		[89]
	Acidithiobacillus ferrooxidans		[87]
	Sulfobacillus thermosulfidooxidans and unidentified acidophilus bacteria (*code A1TSB*)		[90]
	Acidithiobacillus thiooxidans and *Acidithiobacillus ferrooxidans*		[79]
	Chromobacterium violaceum, *Pseudomonas fluorescens* and *Bacillus megaterium*		[91]
PCB for cell phones	*Chromobacterium violaceum*		[61]
PCB	*Acidithiobacillus ferrooxidans* and *Acidithiobacillus thiooxidans*		[92]
PCB TV	*Acidithiobacillus ferrooxidans*, *Leptospirillum ferrooxidans* and *Acidithiobacillus thiooxidans*		[64]
WEEE (dust generated in crushing)	*Acidithiobacillus thiooxidans*, *Acidithiobacillus ferrooxidans*, *Aspergillus niger* and *Penicillium simplicissimum*		[93]
WEEE	*Acidithiobacillus* sp. and *Leptospirillum* sp.		[60]

Bioleaching for treating electrical and electronic waste is still not being employed commercially, but there is an integrated project in Europe for developing the biotechnology (BioMinE), the objective of which is to provide European countries with a strong research and development structure through encouraging joint work between research laboratories and end users. This should therefore favor the appearance of innovative biohydrometallurgical processes, not only for ore, but also for secondary sources [46].

Recent legislation in several countries has driven research into the treatment and recovery of metals from electrical and electronic waste, but the studies are still very small in scale.

Nevertheless, there appears to be a promising future for bioleaching as far as the recovery of metals from secondary sources is concerned, for its application is economically, environmentally and legally favorable.

References

1. Van der Sloot HA, Heasman L, Quevauviller PH (1996) Conclusions of the workshop: harmonization of leaching/extraction tests for environmental risk assessment. Sci Total Environ 178(1–3):133–139
2. Geankoplis CJ (1993) Transport process and unit operations, 3rd edn. Prentice-Hall International Inc., New Jersey
3. Richardson JF, Harker JH, Backhurst JR (2002) Coulson and Richardson's—chemical engineering: particle technology and separation processes, vol 2. Butterworth-Heinemann, Elsevier Science, Oxford
4. Sohn HY, Wadsworth ME (1979) Rate processes of extractive metallurgy. Plenum Press, New York
5. Cui J, Zhang L (2008) Metallurgical recovery of metals from electronic waste: a review. J Hazard Mater 158:228–256
6. Tuncuk A, Stazi V, Akcil A, Yazici EY, Deveci H (2012) Aqueous metal recovery techniques from e-scrap: hydrometallurgy in recycling. Miner Eng 25:28–37
7. Jergensen GV (1999) Copper leaching; solvent extraction, and electrowinning technology. Society for Mining, Metallurgy, and Exploration Inc., Colorado
8. Ghorbani Y, Becker M, Mainza A, Franzidis J, Petersen J (2011) Large particle effects in chemical/biochemical heap leach processes—a review. Miner Eng 24:1172–1184
9. Syed S (2012) Recovery of gold from secondary sources—a review. Hydrometallurgy 115–116:30–51
10. Park YJ, Fray DJ (2009) Recovery of high purity precious metals from printed circuit boards. J Hazard Mater 164:1152–1158
11. Zhang Y, Liu S, Xie H, Zeng X, Li J (2012) The 7th international conference on waste management and technology—current status on leaching precious metals from waste printed circuit boards. Proc Environ Sci 16:560–568
12. Birloaga I, Michelis I, Ferella F, Buzatu M, Vegliò F (2013) Study on the influence of various factors in the hydrometallurgical processing of waste printed circuit boards for copper and gold recovery. Waste Manag 33:935–941
13. Ha VH, Lee J, Jeong J, Hai HT, Jah MK (2010) Short communication—thiosulfate leaching of gold from waste mobile phones. J Hazard Mater 178:1115–1119
14. Petter PMH, Veit HM, Bernardes AM (2014) Evaluation of gold and silver leaching from printed circuit board of cellphones. Waste Manag 34:475–482
15. Yamane LH, Moraes VT, Espinosa DCR, Tenório JAS (2011) Recycling of WEEE: characterization of spent printed circuit boards from mobile phones and computers. Waste Manag 31:2553–2558
16. Veit HM, Bernardes AM, Ferreira JZ, Tenório JAS, Malfatti CF (2006) Recovery of copper from printed circuit boards scraps by mechanical processing and electrometallurgy. J Hazard Mater B137:1704–1709

17. Zeng X, Zheng L, Xie H, Lu B, Xia K, Chao K, Li W, Yang J, Lin S, Li J (2012) The seventh international conference on waste management and technology—current status and future perspective of waste printed circuit boards recycling. Proc Environ Sci 16:590–597
18. Behnamfard A, Salarirad MM, Veglio F (2013) Process development for recovery of copper and precious metals from waste printed circuit boards with emphasize on palladium and gold leaching and precipitation. Waste Manag 33(11):2354–2363
19. Castro LA, Martins AH (2009) Recovery of tin and copper by recycling of printed circuit boards from obsolete computers. Braz J Chem Eng 26(4):649–657
20. Oh CJ, Lee SO, Yang HS, Ha TJ, Kim MJ (2003) Selective leaching of valuable metals from waste printed circuit boards. J Air Waste Manag Assoc 53:897–902
21. Yazici EY, Deveci H (2013) Extraction of metals from waste printed circuit boards (WPCBs) in H_2SO_4–$CuSO_4$–NaCl solutions. Hydrometallurgy 139:30–38
22. Jah MK, Kumari A, Choubey PK, Lee J, Kumar V, Jeong J (2012) Leaching of lead from solder material of waste printed circuit boards (PCBs). Hydrometallurgy 121–124:28–34
23. Xu J, Thomas HR, Francis RW, Lum KR, Wang J, Liang B (2008) Review—a review of processes and technologies for the recycling of lithium-ion secondary batteries. J Power Sources 177:512–527
24. Zhu S, He W, Li G, Zhou X, Zhang X, Huang J (2012) Recovery of Co and Li from spent lithium-ion batteries by combination method of acid leaching and chemical precipitation. Trans Nonferrous Met Soc China 22:2274–2281
25. Shin SM, Kim NH, Sohn JS, Yang DH, Kim YH (2005) Development of a metal recovery process from Li-ion battery wastes. Hydrometallurgy 79:172–181
26. Swain B, Jeong J, Lee J, Lee G, Sohn J (2007) Hydrometallurgical process for recovery of cobalt from waste cathodic active material generated during manufacturing of lithium ion batteries. J Power Sources 167:536–544
27. Chen L, Tang X, Zhang Y, Li L, Zeng Z, Zhang Y (2011) Process for the recovery of cobalt oxalate from spent lithium-ion batteries. Hydrometallurgy 108:80–86
28. Jah MK, Kumari A, Jah AK, Kumar V, Hait J, Pandey BD (2013) Recovery of lithium and cobalt from waste lithium ion batteries of mobile phone. Waste Manag 33(9):1890–1897
29. Méar F, Yot P, Cambon M, Ribes M (2006) The characterization of waste cathode-ray tube glass. Waste Manag 26:1468–1476
30. Musson SE, Jang YC, Townsend TG, Chung IH (2000) Characterization of lead leachability from cathode ray tubes using the toxicity characteristic leaching procedure. Environ Sci Technol 34:4376–4381
31. Pruksathorn K, Damronglerd S (2005) Lead recovery from waste frit glass residue of electronic plant by chemical-electrochemical methods. Korean J Chem Eng 22:873–876
32. Herrero M, Mendiola JA, Cifuentes A, Ibáñez E (2009) Review: supercritical fluid extraction: recent advances and applications. J Chromatogr A 1217(16):2495–2511
33. Brunner G (2005) Supercritical fluids: technology and application to food processing. J Food Eng 67:21–33
34. Erkey C (2000) Supercritical carbon dioxide extraction of metals from aqueous solutions: a review. J Supercrit Fluids 17:259–287
35. Laurent A, Lack E, Gamse T, Marr R (2001) High pressure process technology: fundamentals and applications. Ind Chem Libr 9 (Elsevier)
36. Chen Y, Mariba ERM, Van Dyk L, Potgieter JH (2011) A review of non-conventional metals extracting technologies from ore and waste. Int J Miner Process 98:1–7
37. Sunarso J, Ismadji S (2009) Decontamination of hazardous substances from solid matrices and liquids using supercritical fluids extraction: a review. J Hazard Mater 161:1–20
38. Pourmortazavi SM, Hajimirsadeghi SS, Kohsari I, Hosseini SG (2004) Orthogonal array design for the optimization of supercritical carbon dioxide extraction of different metals from a solid matrix with Cyanex 301 as a ligand. J Chem Eng Data 49:1530–1534
39. Yamini Y, Saleh A, Khajeh M (2008) Orthogonal array design for the optimization of supercritical carbon dioxide extraction of platinum(IV) and rhenium(VII) from a solid matrix using Cyanex 301. Sep Purif Technol 61:109–114

40. Wai CM, Wang S (1997) Review—supercritical fluid extraction: metals as complexes. J Chromatogr A 785:369–383
41. Xiu F, Qi Y, Zhang F (2013) Recovery of metals from waste printed circuit boards by supercritical water pre-treatment combined with acid leaching process. Waste Manag 33:1251–1257
42. Xiu F, Zhang F (2009) Recovery of copper and lead from waste printed circuit boards by supercritical water oxidation combined with electrokinetic process. J Hazard Mater 165:1002–1007
43. Zhang F, Xiu F (2010) Materials recovery from waste printed circuit boards by supercritical methanol. J Hazard Mater 178:628–634
44. Brandl H (2001) Microbial Leaching of Metals. In: Rehm H-J, Reed G (eds) Biotechnology Set, 2nd edn. Wiley-VCH, pp 191–224
45. Rossi G (1990) Historical background. Biohydrometallurgy. McGraw-Hill, pp 1–13
46. Morin D (2006) BioMinE—integrated project for the development of biotechnology for metal-bearing materials in Europe. Hydrometallurgy 83(1–4):69–76
47. Ehrlich HL (1997) Microbes and metals. Appl Microbiol Biotechnol 48(6):687–692
48. Brierley CL, Brierley JA (2013) Progress in bioleaching: part B: applications of microbial processes by the minerals industries. Appl Microbiol Biotechnol 97(17):7543–7552
49. Bayat B, Sari B (2010) Comparative evaluation of microbial and chemical leaching processes for heavy metal removal from dewatered metal plating sludge. J Hazard Mater 174(1–3):763–769
50. Wang Q (2009) Effects of water-washing pretreatment on bioleaching of heavy metals from municipal solid waste incinerator fly ash. J Hazard Mater 162(2–3):812–818
51. Yang J (2009) Biosorption behavior of heavy metals in bioleaching process of MSWI fly ash by *Aspergillus niger*. Biochem Eng J 46(3):294–299
52. Xu TJ, Ting YP (2009) Fungal bioleaching of incineration fly ash: metal extraction and modeling growth kinetics. Enzyme Microb Technol 44(5):323–328
53. Seidel H (2006) Effect of different types of elemental sulfur on bioleaching of heavy metals from contaminated sediments. Chemosphere 62(9):1444–1453
54. Zhu N (2003) Recycling of spent nickel–cadmium batteries based on bioleaching process. Waste Manag 23(8):703–708
55. Zhao L, Yang D, Zhu NW (2008) Bioleaching of spent Ni–Cd batteries by continuous flow system: Effect of hydraulic retention time and process load. J Hazard Mater 160(2–3):648–654
56. Xin B (2009) Bioleaching mechanism of Co and Li from spent lithium-ion battery by the mixed culture of acidophilic sulfur-oxidizing and iron-oxidizing bacteria. Bioresour Technol 100(24):6163–6169
57. Xin B (2012) Bioleaching of zinc and manganese from spent Zn–Mn batteries and mechanism exploration. Bioresour Technol 106:147–153
58. Mishra D (2008) Bioleaching of metals from spent lithium ion secondary batteries using *Acidithiobacillus ferrooxidans*. Waste Manag 28(2):333–338
59. Zhu N (2011) Bioleaching of metal concentrates of waste printed circuit boards by mixed culture of acidophilic bacteria. J Hazard Mater 192(2):614–619
60. Vestola EA (2010) Acid bioleaching of solid waste materials from copper, steel and recycling industries. Hydrometallurgy 103(1–4):74–79
61. Chi TD (2011) Bioleaching of gold and copper from waste mobile phone PCBs by using a cyanogenic bacterium. Miner Eng 24(11):1219–1222
62. Xiang Y (2010) Bioleaching of copper from waste printed circuit boards by bacterial consortium enriched from acid mine drainage. J Hazard Mater 184(1–3):812–818
63. Ilyas S, Lee JC, Chi RA (2013) Bioleaching of metals from electronic scrap and its potential for commercial exploitation. Hydrometallurgy 131–132:138–143
64. Bas AD, Deveci H, Yazici EY (2013) Bioleaching of copper from low grade scrap TV circuit boards using mesophilic bacteria. Hydrometallurgy 138:65–70

65. Ilyas S (2010) Column bioleaching of metals from electronic scrap. Hydrometallurgy 101(3–4):135–140
66. Asghari I (2013) Bioleaching of spent refinery catalysts: a review. J Ind Eng Chem 19(4):1069–1081
67. Amiri F (2012) Bioleaching kinetics of a spent refinery catalyst using *Aspergillus niger* at optimal conditions. Biochem Eng J 67:208–217
68. Gholami RM, Borghei SM, Mousavi SM (2011) Bacterial leaching of a spent Mo–Co–Ni refinery catalyst using *Acidithiobacillus ferrooxidans* and *Acidithiobacillus thiooxidans*. Hydrometallurgy 106(1–2):26–31
69. Lewis G (2011) Bio hydrometallurgical recovery of metals from fine shredder residues. Miner Eng 24(11):1166–1171
70. Qu Y (2013) Bioleaching of heavy metals from red mud using *Aspergillus niger*. Hydrometallurgy 136:71–77
71. Qu Y, Lian B (2013) Bioleaching of rare earth and radioactive elements from red mud using Penicillium tricolor RM-10. Bioresour Technol 136:16–23
72. Kaksonen AH (2011) Bioleaching and recovery of metals from final slag waste of the copper smelting industry. Miner Eng 24(11):1113–1121
73. Ren WX (2009) Biological leaching of heavy metals from a contaminated soil by *Aspergillus niger*. J Hazard Mater 167(1–3):164–169
74. Deng X (2012) Bioleaching of heavy metals from a contaminated soil using indigenous *Penicillium chrysogenum* strain F1. J Hazard Mater 233–234:25–32
75. Saidan M, Brown B, Valix M (2012) Leaching of electronic waste using biometabolised acids. Chin J Chem Eng 20(3):530–534
76. Tuncuk A (2012) Aqueous metal recovery techniques from e-scrap: hydrometallurgy in recycling. Miner Eng 25(1):28–37
77. Mousavi SM, Yaghmaei S, Salimi F, Jafari A (2006) Influence of process variables on biooxidation of ferrous sulfate by na indigenous Acidithiobacillus ferrooxidans. Part I: flask experiments. Fuel 85:2555–2560
78. Nemati M, Harrison STL, Hansford GS, Webb C (1998) Biological oxidation of ferrous sulfate by Thiobacillus ferrooxidans: a review on the kinetic aspects. Biochem Engg J 1:171–190
79. Liang G (2013) Optimizing mixed culture of two acidophiles to improve copper recovery from printed circuit boards (PCBs). J Hazard Mater 250–251:238–245
80. Giaveno MA, Urbieta MS, Donati E (2011) Mechanisms of bioleaching: basic understanding and possible industrial implications. Biohydrometallurgical process: a practical approach. CETEM, MCTI, Rio de janeiro, pp 24–38
81. Sand W, Gehrke T, Jozsa P, Schippers A (2001) (Bio)chemistry of bacterial leaching—direct vs. Indirect bioleaching. Hydrometallurgy 59:159–175
82. Rohwerder T, Sand W (2007) Mechanisms and biochemical fundamentals of bacterial metal sulfide oxidation. In: Donati ER, Sand W (eds) Microbial Processing of Metal Sulfides. Springer, Netherlands, pp 35–58
83. Schippers A (2007) Microorganisms Involved in Bioleaching and Nucleic Acid-Based Molecular Methods for Their Identification and Quantification. In: Donati ER, Sand W (eds) Microbial Processing of Metal Sulfides. Springer, Netherlands, pp 3–33
84. Lee JC, Pandey BD (2012) Bio-processing of solid wastes and secondary resources for metal extraction—a review. Waste Manag 32(1):3–18
85. Welch SA, Barker WW, Banfield JF (1999) Microbial extracellular polysaccharides and plagioclase dissolution. Geochim Cosmochim Acta 63(9):1405–1419
86. Pant D (2012) Chemical and biological extraction of metals present in E waste: a hybrid technology. Waste Manag 32(5):979–990
87. Choi MS (2004) Microbial recovery of copper from printed circuit boards of waste computer by *Acidithiobacillus ferrooxidans*. J Environ Sci Health A Tox Hazard Subst Environ Eng 39:2973–2982

88. Cerruti C, Curutchet G, Donati E (1998) Bio-dissolution of spent nickel–cadmium batteries using *Thiobacillus ferrooxidans*. J Biotechnol 62(3):209–219
89. Yang T (2009) Factors influencing bioleaching copper from waste printed circuit boards by *Acidithiobacillus ferrooxidans*. Hydrometallurgy 97(1–2):29–32
90. Ilyas S (2007) Bioleaching of metals from electronic scrap by moderately thermophilic acidophilic bacteria. Hydrometallurgy 88(1–4):180–188
91. Faramarzi MA, Stagars M, Pensini E, Krebs W, Brandl H (2004) Metal solubilization from metal-containing solid materials by cyanogenic *Chromobacterium violaceum*. J Biotechnol 113:321–326
92. Wang J (2009) Bioleaching of metals from printed wire boards by *Acidithiobacillus ferrooxidans* and *Acidithiobacillus thiooxidans* and their mixture. J Hazard Mater 172(2–3):1100–1105
93. Brandl H, Bosshard R, Wegmann M (2001) Computer-munching microbes: metal leaching from electronic scrap by bacteria and fungi. Hydrometallurgy 59(2–3):319–326

Chapter 6
Hydrometallurgical Processing

Daniel Assumpção Bertuol, Eduardo Hiromitsu Tanabe, Lucas Meili and Hugo Marcelo Veit

Hydrometallurgical processes are used to recover metals from solutions obtained after leaching steps. This chapter will present liquid-liquid extraction and cementation processes used to recover metals from WEEE.

6.1 Liquid-Liquid Extraction

Liquid-liquid extraction, sometimes called solvent extraction, is the separation of the constituents of a liquid solution by contact with another insoluble liquid [1]. The process enables the separation of the constituents of an aqueous solution by contact with another insoluble liquid, an organic solution. Through intensive contact, the solute transfers from the aqueous solution (raffinate) into the organic solvent (extract). After mixing, the two phases are separated, which happens either by gravity or by centrifugal forces. To recover the solvent and to obtain the solute in pure form, a further separation process is necessary (rectification or re-extraction).

D.A. Bertuol (✉) · E.H. Tanabe
Laboratory of Environmental Process (LAPAM), Center of Technology UFSM,
Santa Maria, Brazil
e-mail: dbertuol@gmail.com

E.H. Tanabe
e-mail: edutanabe@ufsm.br

L. Meili
Laboratory of Separations Systems and Process Optimization (LASSOP),
Center of Technology—UFAL, Maceió, Brazil
e-mail: lucasmeili@gmail.com

H.M. Veit
Programa de Pós-Graduação em Engenharia de Minas, Metalúrgica e de Materiais—PPGE3M, Universidade Federal do Rio Grande do Sul—UFRGS, Porto Alegre,
Rio Grande do Sul, Brazil
e-mail: hugo.veit@ufrgs.br

H.M. Veit and A. Moura Bernardes (eds.), *Electronic Waste*, Topics in Mining, Metallurgy and Materials Engineering, DOI 10.1007/978-3-319-15714-6_6

In the ideal case, the component to be extracted is soluble in the solvent and the other components are insoluble. This means the solute is the only component of the initial mixture transferred to the solvent phase. The initial mixture becomes a raffinate as the solute is extracted. The solvent phase becomes the extract as it receives the solute. In practice, all components are potentially soluble to some degree in each other, and the separation is only feasible when the solubilities are sufficiently different. In any case, the component that is not extracted (inert) must be sufficiently insoluble so as to form two phases that can be separated [2].

The separation of a component from a homogeneous solution is obtained by the addition of another insoluble constituent, the solvent, in which the desired component, the solute, is preferably soluble. The solute diffuses with a characteristic velocity until equilibrium concentrations are reached in each stage. For example, zinc and cadmium are extracted from the selectively mixed electrolyte solutions containing zinc, cadmium and cobalt sulphates, using di-2-ethylhexyl phosphoric acid (DEHPA) as extractant [3].

Liquid-liquid extraction is used to remove undesirable components from lubricating oils and other fractions from crude oil to produce concentrated phosphoric acid, to separate/concentrate metals and rare earth elements, and in many other applications. Many metal separations, particularly those that are expensive through chemical routes, such as uranium-vanadium, niobium-tantalum, and hafnium-zirconium, can be done economically by extraction [1, 2].

Solvent extraction is an equilibrium process:

$$M^{n+}_{aq} +_n HR \leftrightarrow MR_n +_n H^{+}_{aq} \tag{6.1}$$

where HR is the organic extractant, MR_n is the metal-organic extracted species, H^+ is the proton released by the organic extractant in exchange for the cationic metal species M_n^+. The distribution of the metal ions between the organic and aqueous phases depends on a number of variables, such as solution composition, extraction/equilibrium pH, extractant concentration and type, aqueous to organic volume ratio and, sometimes, on the type of diluent used to dilute the extractant [3].

Liquid-liquid extraction operations may be carried out either as a batch or as a continuous process. In the single-stage batch process illustrated in Fig. 6.1, the solvent and solution are mixed together and then allowed to separate into the two phases—the extract E containing the required solute in the added solvent and the raffinate R, the weaker solution with some associated solvent. With this simple arrangement, mixing and separation can occur in the same vessel [4].

An example of a countercurrent extraction in a series of mixing and separating vessels is shown in Fig. 6.2. Each box corresponds to a mixer and a separator. The initial solution of the solute B in solvent A, F_0, is fed into the first unit and leaves as raffinate F_1. This stream passes through the units and leaves from the *N*th unit as stream F_N. The fresh solvent O_{N+1} enters the *N*th unit and passes in the reverse direction through the units, leaving as extract O_1 [4].

According to Geankoplis [5], if the components B and C are immiscible in each other and the solvent current O_{N+1} contains the components A and C, and if A

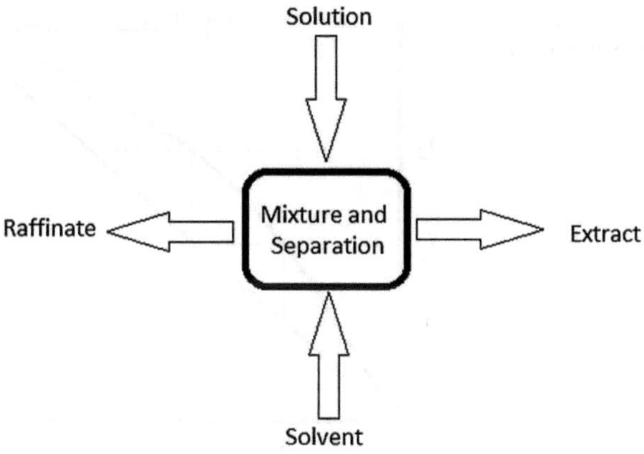

Fig. 6.1 Single-stage batch extraction

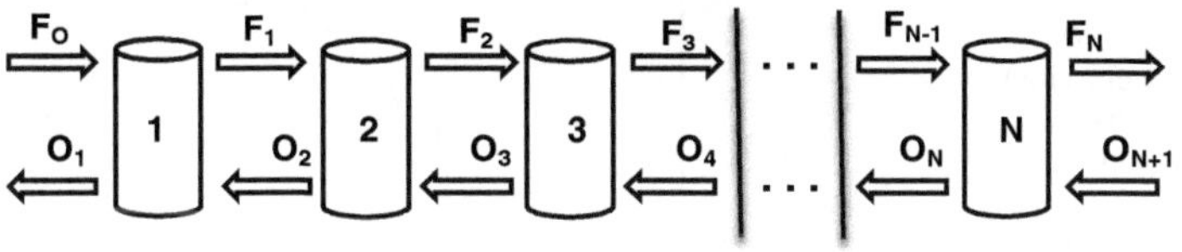

Fig. 6.2 Arrangement for multiple-contact extraction in a countercurrent flow

and B are contained in the feed stream F_0, then the number of stages can be determined. In this configuration, therefore, the solute A is transferred from F_0 to O_{N+1}. According to these considerations, a mass balance for the system (Fig. 6.2) can be done to the first n stages:

$$F'\left(\frac{x_0}{1-x_0}\right)+O'\left(\frac{y_{n+1}}{1-y_{n+1}}\right)=F'\left(\frac{x_n}{1-x_n}\right)+O'\left(\frac{y_1}{1-y_1}\right) \tag{6.2}$$

F′ and O′ represent the mass of the inert components B and C per unit of time, respectively, y is the mass fraction of A in stream O and x is the mass fraction of A in stream F. The obtained equation is the operating line. The operating line can be a straight line if y and x are in low concentrations. Therefore, the number of equilibrium stages can be obtained as shown in Fig. 6.3.

The pH has an important effect on the extractions. Figure 6.4 shows the effect of pH on the extractions of zinc, cobalt and cadmium from a synthetic solution. One can see that negligible cobalt and cadmium co-extractions take place at pH < 3. Thus, at an equilibrium pH < 3, zinc could be extracted from the mixed electrolyte solution with little or no contamination from the other metals. At 3 < pH < 4.5, cadmium extraction is significant but co-extraction of cobalt is negligible. The implication of this observation is that by carefully controlling the extraction/equilibrium pH, a solution containing these electrolytes can be separated into their respective fractions [3].

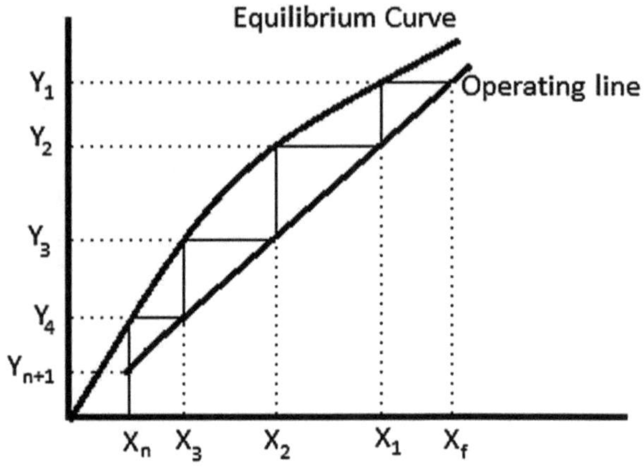

Fig. 6.3 Graphical method for determining the number of stages using immiscible solvents [4]

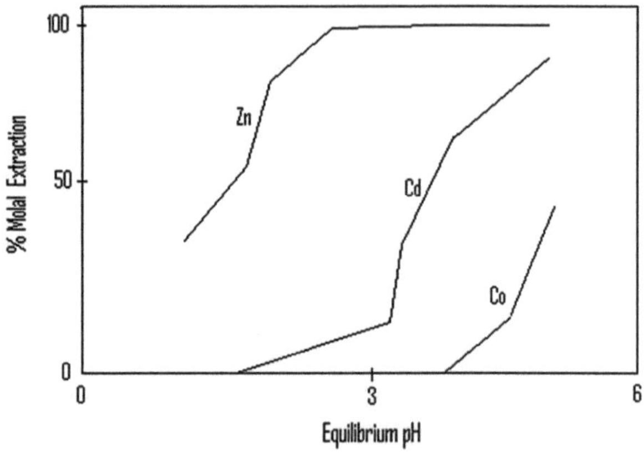

Fig. 6.4 Extent of metal extraction (qualitative behavior): [DEHPA] = 30 v/o; [TBP] = 4 v/o; diluent—SX-1; A: O = 1; temperature = 25 °C; feed solution—21 g/l Zn, 12.6 g/l Cd and 2 g/l Co. (adapted from Owusu [3])

Obtaining metals from aqueous solutions is a technological challenge. It is very difficult to separate the metals among themselves and from other industrial metals present in the leached solutions. Among the various techniques of separation or purification in hydrometallurgy, solvent extraction has great versatility for the separation of different types of elements, and can occur in a wide range of concentrations of solutions. As such, this technique has been widely used for the treatment of leachate concentrates [6].

Extractants are divided into three classes: acidic, basic and neutral or anionic. The difference between them is the chemistry of their reactions, but they must always present specific features for effective extraction. Some of the most important features are: selectivity, high extraction capacity, stability, reversibility and low cost. In general, extractants do not have all these characteristics, so they may be diluted with other components in order to provide the most suitable physical properties (viscosity, density, etc.) and increase their extraction potential and selectivity. The mixture of extractants is a very interesting way to improve extraction efficiency and selectivity, this procedure is called synergistic extraction [7–9].

The application of solvent extraction is important for the purification of solutions in the battery recycling process, and for the recovery of metals of economic importance. The studies presented next have evaluated the liquid-liquid extraction of metals using different extractants.

Devi et al. [10] studied the liquid-liquid extraction of Zn and Mn from sulphate solutions using Cyanex 272. They observed that the extraction becomes more efficient with increasing pH and by increasing the concentration of the extractant to 0.05 M. Kongolo et al. [11] studied the recovery of Co and Zn from a copper sulfate solution by solvent extraction. The use of LIX 984 allowed the recovery of Cu. After $CaCO_3$ precipitation. Fe, Zn and Co were extracted with DEHPA. They obtained high recovery percentages of 95 % for Cu and 90 % for Co and Zn.

Safarzadeh et al. [12] evaluated the solvent extraction of Cd with the following extractors: D2EHPA, Cyanex 272, TBP, TOPS99, PC88A, DIPSA, TIBPS, HRJ-4277 and BTMHA. They found that the extractor D2EHPA was more widespread inapplications in industrial solutions for the recovery of Zn and Cd with pH 1–2. They also found from the equilibrium curves with carboxylic acids (Versatic and naphthenic acids) for Zn and Cd, that the extraction was possible only with pH above 6.5 [12].

Parus et al. [13] studied solvent extraction of Cd (II) from chloride solutions using [1-(2-pyridyl)-tridecane-1-one (2PC12)], [1-(2-pyridyl)-pentadecane-1-one (2PC14)], [1-(4-pyridyl)-tridecane-1-one (4PC12)] and [1-(4-pyridyl)-pentadecane-1-one (4PC14)] oximes. The influence of the extractant concentration, metal concentration, concentrations of chloride ions and various polar and non-polar solvents was evaluated. They found that only the 2-pyridyl ketoximes extracted Cd (II) and that the extraction depends on the solvent, on the reagent concentration and on the concentration of chloride ions. Cd (II) was extracted efficiently by using a mixture of chloroform and hydrocarbons as diluents with decan-1-ol, and removed from the loaded organic phase with aqueous ammonia and water.

Nan et al. [14] studied a new process for recovering metals from a mixture of worn lithium-ion and nickel-metal hydride (NiMH) batteries. In the developed process, liquid-liquid extraction was used to remove copper, cobalt and nickel from the obtained leachate. Using the developed process, it was possible to recover more than 94 % of all metals.

Innocenzi and Veglio [15] studied the use of solvent extraction for purifying leach solutions of NiMH batteries. Dissolved nickel was separated from manganese and zinc by liquid-liquid extraction using DEHPA. Two extractants were investigated: DEHPA and Cyanex 272 in n-dodecane. DEHPA was more efficient in separating manganese and zinc from the leach solutions. The results suggest that a two-step liquid-liquid cross flow system is sufficient to extract 100 % of zinc and about 95 % of manganese, while residual nickel is about 80 % of the initial content. These researchers have proposed a process flow that includes two stages of extraction, 20 % v/v DEHPA in n-dodecane (1/1, at room temperature, 30 min of contact, pH $\leq$ 2.5) and 4 M H_2SO_4 (1/0.5, at room temperature, 15 min of contact).

Larsson et al. [16] studied the properties of extraction with the leaching liquors of nickel metal hydride dissolved in hydrochloric acid 8 M using Cyanex 923.

The obtainment and recovery of rare earth elements (REE) has been studied extensively in recent years. These elements are very relevant from an industrial point of view [17]. REE took on essential importance because of the rapidly growing demand for them in several areas, such as electronics, metallurgy, magnetism, catalysts, ceramics and laser technology. Although there are many different uses for REE, they are not recycled in large quantities. This could change, however, if recycling became mandatory or if very high prices of rare earth metals (REM) made recycling feasible. In-plant recycling activities have been reported by the NEOMAX group (Hitachi Metals Ltda), including its plants in Japan. This process includes a solvent extraction step to recover REE from permanent magnet scraps. Recently, an hydrometallurgical process was developed by Rhodia (France) for the recycling of rare earth oxides (REO) from slag generated after the pyrometallurgical treatment of NiMH batteries. This recovery of REO is a business secret, but the process includes different solvent extraction units [18].

The reported separation processes of REE are based on the use of the following classes of extractants: tri-n-butilfosfate (TBP), quaternary amines, acid versatic and acid di-2-etilhexilfosforic (DEHPA).

Banda et al. [19] developed a process for the separation of Pr from La and Nd in a chloride solution. Among the extractants (Cyanex 272, DEHPA, PC88A, and Cyanex 301), Cyanex 272 and saponified Cyanex 272 showed a better potential for extracting La than Pr and Nd.

Alstad et al. [20] studied the solvent extraction of rare earth metal ions (La, Pr, Nd, Sm, Eu, Tb, Dy, Ho, Er, Tm, Yb and Lu) from an ionic aqueous solution of sodium perchlorate, using carbon tetrachloride as the organic phase. Tian et al. [21] also conducted a study of the synergistic extraction of Sm (III) in a chloride medium. Nasab et al. [22] evaluated the optimal process conditions for the separation of lanthanum, cerium and yttrium using three acids, Cyanex 272, Cyanex 302 and TBP (HA), at concentrations ranging from 0.01 to 0.5 M. The use of Cyanex 272 was more effective for the separation of thorium and rare earths when compared with TBP.

Torkoman et al. [23] conducted studies on the extraction of Sm (III) from nitrate in aqueous solutions, using Cyanex 301 alone and in combination with DEHPA. They evaluated the influence of various operating parameters, such as extraction time, acidity of the aqueous solution, concentration of the extractor and temperature. The synergistic effect was also evaluated, and the addition of DEHPA to Cyanex 301 in low concentrations (0.01–0.05 M) increased the efficiency of the separation.

Tian et al. [21] also studied the synergistic solvent extraction of rare earth elements (La, Nd, Sm, Tb, Ho, Tm) from nitrate mixtures with 8-hydroxyquinoline (HQ) and acidic organophosphorus extractants, Cyanex 301 and Cyanex 302.

Catalysts are widely used in various operations in the petroleum refining and petrochemical industry. These catalysts undergo a process of deactivation due to coke deposits. They are therefore periodically regenerated and replaced by fresh catalysts. However, when the regeneration process is no longer possible, catalysts are treated for the recovery of the noble metals present in their composition [24, 25]. According to Li et al. [26] spent catalysts are dangerous waste, but they are valuable secondary materials. This has motivated companies and research groups to develop efficient methods for regenerating and reusing saturated or deactivated catalysts.

Chen et al. [27] studied the extraction of molybdenum and vanadium from the ammonia leaching residue of spent catalysts by roasting the residue with soda carbonate, followed by a hydrometallurgical treatment of the roasted products. The researchers found that the proposed roasting, leaching and separation steps provide a feasible alternative for the processing of ammonia leaching residue of spent catalysts, and that it can be applied in the comprehensive utilization of low grade molybdenum ore.

Marinho et al. [25] evaluated a hydrometallurgical route for processing spent commercial catalysts (Pt and PtSnIn/A2O3) used in Brazilian refineries. They studied the recovery of noble metals with less generation of final waste. They found that Aliquat 336 (15 vol.% in kerosene) extracted more than 99 wt% of platinum in one stage at 25 °C with an A/O phase ratio equal to 1 (v/v). No other metals were extracted. Barakat and Mahmoud [28] studied the recovery of platinum from spent catalysts. Platinum was separated from the leach liquor by direct precipitation and by solvent extraction using trioctylamine in kerosine.

6.2 Supercritical Extraction

The increasing environmental problems caused by the pollution of organic compounds and the restrictions imposed by environmental legislation have limited the use of conventional solvents. In this context, the use of supercritical fluids (SCF) is a promising alternative for the replacement of conventional toxic organic solvents [29].

According to Fox et al. [30], supercritical fluids (SCFs) have been successfully used in industrial-scale extractions since the 1930s–1940s. It is most commonly used in the extraction and separation of organic compounds, and the traditionally employed solvents are Carbon dioxide and light hydrocarbon gases. These authors studied the complex formation reactions of praseodymium nitrate hexahydrate, and neodymium nitrate hexahydrate salts with tri-n-butyl phosphate (TBP) and several other neutral organophosphorus reagents in supercritical carbon dioxide.

Tai et al. [29] studied the kinetics of metal ion (Zinc II as model species) extraction using an in situ chelation-SFE method with Cyanex 302 as the chelating agent and supercritical carbon dioxide as the solvent, which extracts the metal-chelate complex from aqueous solutions. Their main conclusion was that the extraction rate of zinc (II) ions increases with an increase in stirring rate, but decreases with pressure.

Goreishi et al. [31] evaluated the extraction of toxic heavy metals, uranium (U), hafnium (Hf) and zirconium (Zr), using supercritical carbon dioxide (SC-CO_2) and Cyanex 301 as chelating agent from a synthetic wastewater sample. The researchers found that the extraction yield of U, Hf and Zr at optimal conditions was 98.1, 28.9 and 65.2 %, respectively.

Liang et al. [32] studied the extraction of cobalt ions from a fireproof board using a mixture of D2EPHA (Di-2-Ethyl Hexyl Phosphoric Acid) and n-hexane in supercritical carbon dioxide. The authors found that the cobalt on the spiked sample can be totally removed in the evaluated pilot unit.

6.3 Cementation

Cementation is a unit process in hydrometallurgical engineering and it is used to precipitate a metal from its solution onto another metal that is more electropositive. The thermodynamic conditions necessary for cementation can be determined by considering the reduction potentials of the species involved in the reaction. In general, the cementation reaction can be expressed as:

$$M^{n+} + ne = M^0 \tag{6.3}$$

The precipitation of copper on iron from a natural solution is a classical example of a relatively ancient art that has been applied successfully for centuries to produce copper on a commercial scale. According to Lung [33], this method has been applied in China since 1086 AD for the extraction of copper from mine water.

In the case of copper, for example, dilute solutions of copper are put into contact with metallic iron, leading to the general reaction,

$$Cu^{2+}(aq) + Fe^0 = Fe^{2+}(aq) + Cu^0 \tag{6.4}$$

Cementation reactions involve the transfer of electrons and are thus by definition electrochemical reactions. The deposition of copper ions and the transfer of iron ions into the solution can be treated as a corrosion reaction.

Cementation is used extensively as a primary metal recovery method for cadmium, copper, gold and silver, and as an electrolyte purification technique in electrolytic processes.

The most common industrial cementation operations include the use of zinc dust to precipitate gold and silver from cyanide solutions, and the use of iron to recover copper from copper-bearing solutions [33].

Furthermore, several researchers have used the cementation process to concentrate/recover unusual metals. Cao [34] used cementation to obtain cobalt using two different ferromanganese alloys. The results show that the reaction is quick and reached levels close to 100 % recovery of cobalt.

In another study, Safarzadeha et al. [35] evaluated cadmium cementation using zinc dust. The results showed a recovery rate of 95.83 % for cadmium. Anacleto [36] studied Mercury cementation using zinc, iron and metallic aluminium as reducing agents. The reaction efficiency is strongly dependent on pH. For each metal under study, an ideal pH was established. It was shown that mercury cementation with metallic zinc is a first-order process referring to mercury concentration.

Aktas [37] investigated the cementation of rhodium from waste chloride solutions using a metallic copper powder as reducing agent. The method used in this study resulted in a fine rhodium powder with a purity of 98.65 %. The copper content of the powder was found to be 1,220 ppm. The impure powder can be directly sold to any refinery, but it requires further purification prior to being used for the preparation of a rhodium sulfate plating solution.

The recovery of metals, especially gold, silver and copper, from electronic waste with the cementation process has also been extensively investigated by several authors [38–42].

References

1. Treybal RE (1968) Mass transfer operations, 2nd edn. McGraw-Hill, New York
2. Foust AS, Wenzel LA, Clump CW, Maus L, Andersen LB (1980) Principles of unit operations. 2nd edn. Wiley, New York
3. Owusu G (1998) Selective extractions of Zn and Cd from Zn-Cd-Co-Ni sulphate solution using di-2-ethylhexyl phosphoric acid extractant. Hydrometallurgy 47:205–215
4. Harker JH, Backhurst JR, Richardson JF (2002) Chemical engineering, vol 2: particle technology and separation process, vol 5. Elsevier, London
5. Geankoplis CJ (2003) Transport process and separation process principles, 4th edn. Prentice Hall, upper Saddle River
6. Thornton JD (1992) Science and practice of liquid-liquid extraction. Clarendon Press, Oxford
7. Habashi F (1999) Textbook of hydrometallurgy. 2nd edn. Laval University Canada
8. Erkey C (2000) Supercritical carbon dioxide extraction of metals from aqueous solutions: a review. J Supercrit Fluids 17:259–287
9. Tong S, Zhao X, Song N, Jia Q, Zhou W, Liao W (2009) Solvent extraction study of rare earth elements from chloride medium by mixtures of sec-nonylphenoxy acetic with Cyanex301 and Cyanex302. Hydrometallurgy 100:15–19
10. Devi NB, Nathsarma KC, Chakravortty V (1997) Extraction and separation of Mn(II) and Zn(II) from sulphate solutions by sodium salt of Cyanex 272. Hydrometallurgy 45(1–2):169–179
11. Kongolo K, Mwema MD, Banza AN, Gock E (2003) Cobalt and zinc recovery from copper sulphate solution by solvent extraction. Miner Eng 16(12):1371–1374
12. Safarzadeh MS, Bafghi MS, Moradkhani D, Ojaghi IM (2007) A review on hydrometallurgical extraction and recovery of cadmium from various resources. Miner Eng 20:211–220

13. Parus A, Wieszczycka K, Olszanowski A (2011) Solvent extraction of cadmium(II) from chloride solutions by pyridyl ketoximes. Hydrometallurgy 105(3–4):284–289
14. Nan J, Han D, Yang M, Cui M, Hou X (2006) Recovery of metal values from a mixture of spent lithium-ion batteries and nickel-metal hydride batteries. Hydrometallurgy 84(1–2):75–80
15. Innocenzi V, Veglio F (2012) Separation of manganese, zinc and nickel from leaching solution of nickel-metal hydride spent batteries by solvent extraction. Hydrometallurgy 129–130:50–58
16. Larsson K, Ekberg C, Odegaard-Jensen A (2012) Using Cyanex 923 for selective extraction in a high concentration chloride medium on nickel metal hydride battery waste. Hydrometallurgy 129–130:35–42
17. Humphries M (2010) Rare Earth elements: the global supply chain. Congressional Research Service, United States Congress
18. Worrell E, Reuter M (eds) (2014) Handbook of recycling. Elsevier, London
19. Banda R, Nguyen TH, Sohn SH, Lee MS (2013) Recovery of valuable metals and regeneration of acid from the leaching solution of spent HDS catalysts by solvent extraction. Hydrometallurgy 133:161–167
20. Alstad J, Augustson JH, Farbu L (1974) Solvent extraction of rare-earth metal ions with thenoyltrifluoroacetone in carbon tetrachloride. J Inorg Nucl Chem 36(4):899–903
21. Miaomiao T, Qiong J, Wuping L (2013) Studies on synergistic solvent extraction of rare earth elements from nitrate medium by mixtures of 8-hydroxyquinoline with Cyanex 301 or Cyanex 302. J Rare Earths 31(6):604
22. Nasab ME, Sam A, Milani SA (2011) Determination of optimum process conditions for the separation of thorium and rare earth elements by solvent extraction. Hydrometallurgy 106:141–147
23. Torkaman R, Moosavian MA, Torab-Mostaedi M, Safdari J (2013) Solvent extraction of samarium from aqueous nitrate solution by Cyanex301 and D2EHPA. Hydrometallurgy 137:101–107
24. Kar BB, Murthy BVR, Misra VN (2005) Extraction of molybdenum from spent catalyst by salt-roasting. Int J Miner Process 76:143–147
25. Marinho RS, Afonso JC, Cunha JWSD (2010) Recovery of platinum from spent catalysts by liquid-liquid extraction in chloride médium. J Hazard Mater 179:488–494
26. Li W, Peng J, Zhang L, Zhang Z, Li L, Zhang S, Guo S (2008) Pilot-scale extraction of zinc from the spent catalyst of vinyl acetate synthesis by microwave irradiation. Hydrometallurgy 92:79–85
27. Chen Y, Feng Q, Shao Y, Zhang G, Ou L, Lu Y (2006) Investigations on the extraction of molybdenum and vanadium from ammonia leaching residue of spent catalyst. Int J Miner Process 79:42–48
28. Barakat MA, Mahmoud MHH (2004) Recovery of platinum from spent catalyst. Hydrometallurgy 72:179–184
29. Tai CY, You GS, Chen SL (2000) Kinetics study on supercritical fluid extraction of zinc(II) ion from aqueous solutions. J Supercrit Fluids 18:201–212
30. Fox RV, Duane Ball R, Harrington PB, Rollins HW, Jolley JJ, Wai CM (2004) Praseodymium nitrate and neodymium nitrate complexation with organophosphorus reagents in supercritical carbon dioxide solvente. J Supercrit Fluids 31:273–286
31. Ghoreishia SM, Ansaria K, Ghaziaskarb HS (2012) Supercritical extraction oftoxic heavy metals from aqueous waste via Cyanex 301 as chelating agente. J Supercrit Fluids 72:288–297
32. Liang MT, Liang RC, Lin CH, Hsu PJ, Wu LY, Chen HF, Wu YW, Lee WC (2013) Metal extraction of a spiked solid with supercritical carbon dioxide. J Supercrit Fluids 79:324–329
33. Lung TN (1986) The history of copper cementation on iron—the world's first hydrometallurgical process from medieval China. Hydrometallurgy 17(1):113–129
34. Cao Y, Duby P (2001) Cobalt cementation with ferromanganese. Hydrometallurgy 61:195–205

35. Safarzadeha MS, Moradkhanib D, Ilkhchid MO (2007) Determination of the optimum conditions for the cementation of cadmium with zinc powder in sulfate medium. Chem Eng Process 46:1332–1340
36. Anacleto AL, Carvalho JR (1996) Mercury cementation from chloride solutions using iron, zinc and aluminium. Miner Eng 9(4):385–397
37. Aktas S (2012) Cementation of rhodium from waste chloride solutions using copper powder. Int J Miner Process 114–117:100–105
38. Yap CY, Mohamed N (2007) An electrogenerative process for the recovery of gold from cyanide solutions. Chemosphere 67:1502–1510
39. Tuncuka A, Stazi V, Akcil A, Yazici EY, Deveci H (2012) Aqueous metal recovery techniques from e-scrap: hydrometallurgy in recycling. Miner Eng 25:28–37
40. Naseri Joda N, Rashchi F (2012) Recovery of ultra fine grained silver and copper from PC board scraps. Sep Purif Technol 92:36–42
41. Syed S (2012) Recovery of gold from secondary sources—a review. Hydrometallurgy 115–116:30–51
42. Gros F, Baupb S, Aurousseau M (2011) Copper cementation on zinc and iron mixtures: part 1: results on rotating disc electrode. Hydrometallurgy 106:127–133

Chapter 7
Electrometallurgical Processing

Daniel Assumpção Bertuol, Eduardo Hiromitsu Tanabe and Lucas Meili

Electrowinning is an electrochemical process employed to extract metals such as copper, zinc, nickel, manganese and cadmium from their solutions [1]. The electrolytic cell is the basic device used in this process, consisting of an inert anode, such as lead or titanium, and a cathode, placed in an aqueous electrolyte containing the metal solution. The cathode is either a thin plate of pure metal (starter sheet) or a plate made from stainless steel or aluminium (permanent cathode plate). Metal ions pass from the solution and are deposited onto the cathode; gases such as chlorine or oxygen are evolved at the anode [2].

When the solution is acidic, the oxidation of water molecules at the anode is the common counter reaction, which results in the liberation of oxygen bubbles [1].

$$2H_2O_{(1)} \rightarrow O_{2(g)} + 4H^+_{(aq)} + 4e^- \tag{7.1}$$

The generation of oxygen bubbles generally requires a high overpotential and accounts for a considerable percentage of the energy consumption in metal electrowinning tankhouses. In addition, the generated oxygen bubbles burst at the air/solution interface, producing highly acidic droplets that can form an acid mist throughout the tankhouse of the electrowinning plant. This acidic mist not only poses a serious health hazard to the workers, but it causes extreme discomfort to the skin, eyes and respiratory system of tankhouse workers. It is also highly corrosive and results in the corrosion of cathode plates, anode hanger bars, tankhouse

D.A. Bertuol (✉) · E.H. Tanabe
Laboratory of Environmental Process (LAPAM), Center of Technology, UFSM, Santa Maria, Brazil
e-mail: dbertuol@gmail.com

E.H. Tanabe
e-mail: edutanabe@ufsm.br

L. Meili
Laboratory of Separations Systems and Process Optimization (LASSOP), Center of Technology, UFAL, Maceió, Brazil
e-mail: lucasmeili@gmail.com

H.M. Veit and A. Moura Bernardes (eds.), *Electronic Waste*, Topics in Mining, Metallurgy and Materials Engineering, DOI 10.1007/978-3-319-15714-6_7

equipment and building structures [1, 3]. To reduce the formation of this acid mist, many electrowinning tankhouses employ plastic balls or beads [4]. These plastic objects float on the surface of the electrolyte, providing a barrier that prevents the acidic droplets caused by the bursting of oxygen bubbles from becoming airborne [3].

Copper electrowinning is widely used to obtain pure copper cathodes from leached ore. This process can be used as an example to discuss cathodic and anodic reactions aspects and the current efficiency concept.

The electrowinning of copper from a sulfate electrolyte is expressed by Eq. 7.2:

$$CuSO_4 + H_2O \rightarrow Cu + H_2SO_4 + 1/2\,O_2 \tag{7.2}$$

In the process, copper metal (Cu) is deposited, sulfuric acid is generated, and oxygen evolved. It is convenient to discuss the electrowinning chemistry in terms of the cathodic and anodic process.

The cathodic reduction of copper can be written as follows:

$$Cu^{2+} + 2e^- \rightarrow Cu^0 \tag{7.3}$$

At 65 °C, the standard electrode potential for this reaction is 0.34 V. By applying the Nerst equation to the equilibrium shown in Eq. 7.2, the following expression is produced:

$$E = 0.34 + 0.034 \log a_{Cu^{2+}} \tag{7.4}$$

where E is the equilibrium electrode potential and a_{Cu}^{2+} is the activity of the cupric ions in the electrolyte. The equilibrium electrode potential has been measured at about 0.25 V for pure copper in an electrolyte containing 40 g/L Cu^{+2} and 160 g/L H_2SO_4 at 65 °C.

In concentrated electrolytes, the maximum limiting current density (i) is determined by diffusion through the boundary film and is usually expressed as:

$$i = zFD\frac{C_B}{\delta_N} \tag{7.5}$$

where z is the number of electrons transferred, F is the Faraday constant, D is the diffusion coefficient, C_B is the bulk solution concentration, and δ_N is the boundary layer thickness.

The value of the limiting current density is approximately 450 A m^{-2} for a solution containing 45 g/L Cu^{2+}. Most copper electrowinning plants operate at a much lower current density than the limiting value, since deposit structure becomes nodular, porous and even powdery [5].

The fundamental anodic process occurring during the electrowinning of copper is the decomposition of water according to the following reaction:

$$H_2O \rightarrow 2H^+ + 1/2O_2 + 2e^- \tag{7.6}$$

At 65 °C, the standard electrode potential for this reaction is 1.20 V. The Nerst expression for Eq. 7.6 when oxygen is liberated at 1 atm is

$$E = 1.20 + 0.067\ \mathrm{pH} \tag{7.7}$$

And the equilibrium potential is about 1.16 V. The reversible cell potential for the overall copper electrowinning reaction is $E_{cell} = 0.91$ V. In practice, the actual operating voltage is about twice this value (i.e., ~2 V). This difference is attributed to the irreversibility of the system and is explained in terms of electrochemical overvoltages. In electrowinning, the anodic reaction (i.e., oxygen discharge) is highly irreversible. As a result of this irreversibility, high overvoltages are required for the liberation of oxygen from water [5].

When two or more reactions occur simultaneously at an electrode, the number of coulombs of electricity passed corresponds to the sum of the number of equivalents of each reaction. For example, during the deposition of Cu from a solution of cupric nitrate in dilute nitric acid, three cathodic reactions occur: the deposition of Cu (the reduction of cupric ions) and the reduction of both nitrate and hydrogen ions.

The current efficiency (*CE*) is defined as the number of coulombs required for that reaction, Q_j, divided by the total number of coulombs passed, Q_{total}

$$CE = \frac{Q_j}{Q_{total}} \tag{7.8}$$

The number of coulombs required for reaction (Q_j) can be obtained from

$$Q_j = \frac{w \cdot n \cdot F}{A_{wt}} \tag{7.9}$$

where w is the weight of a product of electrolysis, n is the number of electrons involved in the deposition reaction, F is the Faraday constant (F = 96,487 C mol^{-1}) and A_{wt} is the atomic weight of the metal deposited on the cathode.

The total number of coulombs passed during the electrolysis (Q_{total}) will be

$$Q_{total} = I \cdot t \tag{7.10}$$

where I is current in amperes, and t is the elapsed time in seconds.

An alternative equation defining current efficiency is

$$CE = \frac{w_j}{w_{total}} \tag{7.11}$$

where w_j is the weight of the actually deposited metal j and w_{total} is that which would have been deposited if all the current had been used for depositing the metal j [6].

The recovery of metals by electrowinning has proven to be a very efficient technique that is being used in the recovery of various metals, such as copper, nickel and cobalt, among others. A wide variety of metals is employed as cathode, most are a thin plate of pure metal (starter sheet) or a plate made from stainless steel or aluminium (permanent cathode plate). Inert alloys such as lead-0.5 % silver, platinum, titanium coated with platinum/iridium are the most commonly employed as anode. The application of electrowinning in the recovery of the metals present in different types of electronic scrap, such as printed circuit boards, cathode ray tubes, and batteries, among others, has been studied by various authors, who obtained high recovery rates and current efficiencies in the electrowinning of different metals.

Veit et al. [7] studied the recovery of copper from printed circuit boards scrap by mechanical processing and electrometallurgy. The first stage of this work included mechanical processing, such as comminution followed by size, magnetic and electrostatic separation. Through this process it was possible to obtain a concentrated fraction of metals (mainly Cu, Pb and Sn) and another fraction containing polymers and ceramics. In the second stage, the fraction with a concentration of metals was dissolved with acids and treated in an electrochemical process in order to recover the metals separately, especially copper. The cell was set up using a copper plate (5 cm $\times$ 4 cm) as cathode and a platinum plate (5 cm $\times$ 2.5 cm) as anode. The distance between electrodes was 4 cm. The electrowinning experiments were conducted at room temperature ($\pm$20 °C), applying 40 mA cm^{-2} as current density. The results demonstrated a recovery rate of copper above 98 %, allowing this metal to be used in secondary industries.

Kasper et al. [8] recycled some of the components from printed wiring board scrap from mobile phones using mechanical processing, hydrometallurgy and electrometallurgy techniques. At the end of the mechanical processing, a concentrated fraction with metals was dissolved in aqua regia and submitted to electrowinning for the recovery of copper. For this test, the cell was assembled with a copper cathode and a platinum plate as anode. Considering the results obtained in the chemical analysis, the best value for time x current density was obtained for 90 min and a current density of 6 A/dm^2 to recover 92 % of copper from the solution on the cathode.

Xiao et al. [9] recovered copper from complex fine mixture scrap generated by shredding end-of-life information and communications technology products. The tested leaching agents included sulphuric acid, ammonia–ammonium carbonate and ammonia–ammonium sulphate solutions, with or without the addition of hydrogen peroxide. The results show that ammonia leaching has a high selectivity for copper. Direct electrowinning was conducted with ammonia copper leaching solutions. The solution was placed into a single compartment of an electrolysis cell (open to atmosphere), with a 10 cm $\times$ 5 cm copper or aluminium cathode and a lead anode of the same dimensions. Copper with 98.9 % purity was obtained with a current efficiency of 50–70 %, and with lead as major impurity.

Joda and Rashchi [10] recovered ultra fine grained silver and copper from PC board scrap. Nitric acid was used and the effect of the leaching parameters leachate concentration, temperature, time and S/L ratio, were studied and optimized using response surface methodology. After recovering silver from the leach solution, nitric acid was removed by solvent extraction using TBP dissolved in kerosene. Then, aqueous and organic phases were separated by using a separatory funnel. The aqueous phase was used as electrolyte in electrowinning experiments carried out in a 500 mL beaker. Copper and stainless steel plates were used as cathodes and anodes, respectively. Using optimum conditions, 87.30 % Ag and 97.65 % Cu were recovered. At the electrowinning stage, ultrafine grained metallic copper with a particle size of less than 100 nm was obtained.

Safarzadeh and Moradkhani [11] investigated the electrowinning of cadmium in the presence of zinc. Cadmium metal has a wide variety of applications, such as rechargeable nickel–cadmium (NiCd) and silver–cadmium storage batteries. The electrowinning experiments were performed in a mono compartment cubic cell made of Plexiglass® sheets, placed in a thermostatically controlled water bath, equipped with a digitally controlled thermometer (±0.5 °C). A commercial Pb-0.5 % Ag anode alloy sheet was cut into pieces (100 × 150 × 5 mm), which were then welded to an insulated copper wire. These anodes were placed on both sides of the 110 × 160 × 3 mm aluminum (99.5 % Al) cathode, which was masked to give an active area of 121 × 83 mm. Under the optimum working conditions (initial cadmium concentration 120 g/L, current density 80 A/m^2, free acid concentration 98 g/L, and temperature 40 °C), current efficiency (CE) and specific power consumption (SPC) were 98 % and 1.21 kWh/kg, respectively.

Rudnik and Nikiel [12] recovered cadmium and nickel from spent Ni–Cd batteries. The following consecutive stages were proposed: dismantling, leaching in sulphuric acid, cadmium electrowinning, Fe^{3+} removal and nickel electrowinning. The electrowinning of cadmium was carried out at various pH and current densities (100 and 400 A/m^2). Fe^{3+} ions were removed from the bath by hydroxide precipitation. Nickel electrolyses were conducted at a pH of 9 and two current densities (100 and 400 A/m^2). The electrolyses were carried out with a stainless steel plate as the cathode substrate (20 cm^2) and a platinum net as the anode. The amounts of Cd^{2+}, Ni^{2+} and Fe^{3+} ions in the initial solutions (200 cm^3) were assumed as 100 %. The authors calculated that about 70 % and over 91 % of cadmium was recovered at the current densities of 100 and 400 A/m^2, respectively. The maximum recovery of nickel was about 89 % at 400 A/m^2.

Bertuol et al. [13] studied the recovery of nickel and cobalt from NiMH batteries by electrowinning. The effect of different electrowinning parameters, such as boric acid concentration, temperature, current density, and pH, were determined using different synthetic solutions. The optimized operational parameters were applied in an electrowinning test with a solution achieved by leaching the electrodes of NiMH batteries.

The electrowinning tests were performed galvanostatically in a two-compartment cell separated by an anionic membrane (Fig. 7.1). The cathode consisted of polished stainless steel with a 10 cm^2 area and an anode made of titanium coated with platinum/iridium.

The obtained results demonstrate that the use of a two compartment cell was of great worth regarding the current efficiency increment. The use of a membrane reduces the pH variation, reducing the evolution of hydrogen, and therefore increasing the current efficiency. Another important factor is that the use of a membrane reduces pH fluctuation, contributing to a reduction in the consumption of reagents that would otherwise be added to the bath to maintain the pH steady [13].

The optimum conditions in terms of best current efficiency and deposit cohesion for nickel deposition were: 30 g/L of boric acid, 50 °C, 250 $A\ m^{-2}$, and pH 4. After finishing the tests with the synthetic solutions, new tests were performed using a solution obtained through the leaching of the batteries with sulfuric acid,

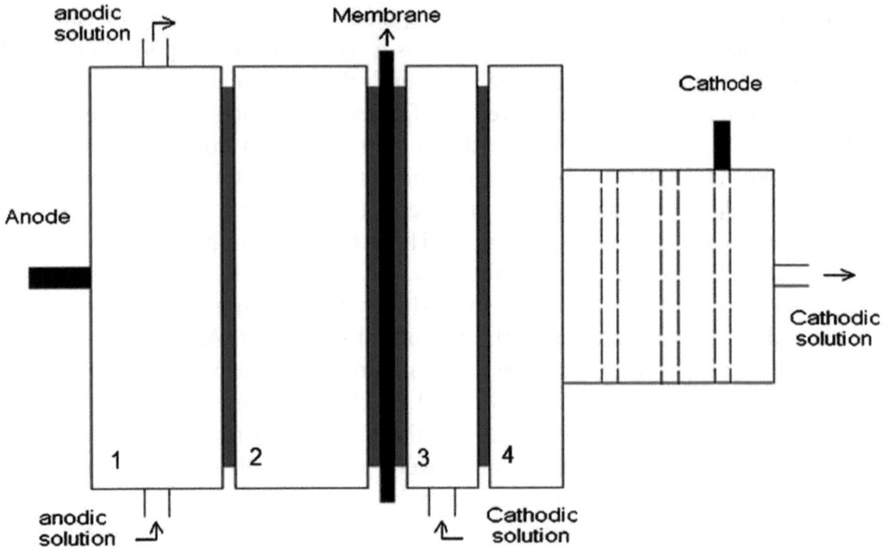

Fig. 7.1 Assembly scheme of two-compartment cell used in the electrowinning tests [13]

after the realization of steps for the precipitation of iron and the removal of rare earth metals. The recovery of nickel and cobalt was successful, since the chemical composition of the deposit presents a significantly high concentration of nickel and cobalt, i.e., 91.92 wt%.

Zhang et al. [14] recovered lead from a cathode ray tube funnel glass by mechano-chemical extraction in an alkaline solution. More than 97 % of Pb in the CRT funnel glass can be extracted with a stirring ball mill leaching process in 5 M sodium hydroxide at 70 °C. The electrowinning of Pb was tested on the leach solution of CRT funnel glasses. The experiments were carried out with the following conditions: initial Pb concentration (17–20 g/L), current densities (400–500 A/m^2) and voltage (1.7–2.0 V). The results showed that a high-purity Pb powder with a metallic Pb concentration of >97 % was deposited in spongy form of a gray color with a slightly bright metallic luster with the electrical current efficiency of 97.5 %.

Industrially, complex end-of-life (EOL) scrap like WEEE are usually integrated in adapted conventional smelter-refinery processes. Copper and lead cycles are mainly used to collect precious metals. In this case, the recovery of precious metals usually includes leaching and electrowinning operations, such as at the Umicore Precious Metals Refining, Hoboken (Belgium) [15].

References

1. Al Shakarji R, He Y, Gregory S (2011) The sizing of oxygen bubbles in copper electrowinning. Hydrometallurgy 109:168–174
2. European Commission (2001) Integrated pollution prevention and control (IPPC). Reference document on best available techniques in the non ferrous metals industries
3. Al Shakarji R, He Y, Gregory S (2013) Performance evaluation of acid mist reduction techniques in copper electrowinning. Hydrometallurgy 131–132:76–80
4. Robinson T, Clayton C (2010) Electrolytic copper electrowinning 2010 world tankhouse survey. Tankhouse Technology Conference, Chile
5. Hiskey JB (1999) Principles and practical considerations of copper electrorefining and electrowinning. Society for Mining, Metallurgy, and Exploration, Inc.
6. Paunovic M (2000) Electrochemical aspects. Wiley, New York
7. Veit HM, Bernardes AM, Ferreira JZ, Tenório JAS, Malfatti CF (2006) Recovery of copper from printed circuit boards scraps by mechanical processing and electrometallurgy. J Hazard Mater B137:1704–1709
8. Kasper AC, Berselli GBT, Freitas BD, Tenório JAS, Bernardes AM, Veit HM (2011) Printed wiring boards for mobile phones: characterization and recycling of copper. Waste Manag 31:2536–2545
9. Xiao Y, Yang Y, Berg JVD, Sietsma J, Agterhuis H, Visser G, Bol D (2013) Hydrometallurgical recovery of copper from complex mixtures of end-of-life shredded ICT products. Hydrometallurgy 140:128–134
10. Joda N, Rashchi F (2012) Recovery of ultra fine grained silver and copper from PC board scraps. Sep Purif Technol 92:36–42
11. Safarzadeh MS, Moradkhani D (2010) The electrowinning of cadmium in the presence of zinc. Hydrometallurgy 105:168–171
12. Rudnik E, Nikiel M (2007) Hydrometallurgical recovery of cadmium and nickel from spent Ni–Cd batteries. Hydrometallurgy 89:61–71
13. Bertuol DA, Amado FDR, Veit H, Ferreira JZ, Bernardes AM (2012) Recovery of nickel and cobalt from spent NiMH batteries by electrowinning. Chem Eng Technol 35(12):2084–2092
14. Zhang C, Wang J, Bai J, Guan J, Wu W, Guo C (2013) Recovering lead from cathode ray tube funnel glass by mechano-chemical extraction in alkaline solution. Waste Manage Res 31(7):759–763
15. Worrell E, Reuter MA (eds) (2014) Handbook of recycling. Elsevier, Amsterdam

Chapter 8
Pyrometallurgical Processing

Denise Crocce Romano Espinosa, Viviane Tavares Moraes and Jorge Alberto Soares Tenório

Conventional metal recycling processes are often based on the primary process routes, meaning that pyrometallurgical processes are applied for the recycling of e.g. copper, lead and steel. Metal casings, external cables and other materials (made of aluminium, steel or copper), separated from WEEE by hand sorting/disassembly, are usually sent to primary or secondary smelters. The process steps depend on the metal content of the secondary raw material, other constituents, size, etc. Nevertheless, for the recycling of metals from complex EOL scrap, such as PCBs, catalysts, LCD, etc., purely hydro- or pyrometallurgical processes are usually not efficient, making a more complex process route necessary.

Since 2010, the EU Raw Materials Initiative has established a list of raw materials at the EU level that are critical in terms of availability. Complex EOL scrap are composite materials and contain some of these critical raw materials, such as Cobalt, Gallium, Germanium, Indium, Platinum Group Metals (PGM) and Rare Earth Elements (REE) [1]. The recycling of these raw materials increases their availability, and is therefore encouraged. Rare metals such as Indium, Galium, REE, and precious metals, including gold, silver and PGM, are usually present in low concentrations in WEEE. The main step for the recovery of these metals usually includes a pretreatment to produce a concentrate that is introduced in conventional extraction processes [2].

Printed circuit boards, processors, connectors, cell phones and MP3 players are processed without any need for grinding or size reduction. The processing of these materials is normally done by a complex route that involves an initial pyrometallurgical

D.C.R. Espinosa (✉) · V.T. Moraes · J.A.S. Tenório
Polytechnic School—USP, São Paulo, Brazil
e-mail: espinosa@usp.br

V.T. Moraes
e-mail: tavares.vivi@gmail.com

J.A.S. Tenório
e-mail: jtenorio@usp.br

H.M. Veit and A. Moura Bernardes (eds.), *Electronic Waste*, Topics in Mining, Metallurgy and Materials Engineering, DOI 10.1007/978-3-319-15714-6_8

stage followed by hydrometallurgical and electrometallurgical operations. There are currently industrial units in Belgium, Germany, Sweden, Canada and Japan [3–5].

Printed circuit boards can contain more than 40 different chemical elements. The value of the scrap depends on the concentration of metals that can be recovered, especially the precious metals, those from the platinum and copper group. The concentration of metals and other materials in printed circuit boards varies a great deal depending on the type of board. Just to give an overall idea, the concentration of metals in a printed circuit board is around 7 % Fe, 5 % Al, 20 % Cu, 1.5 % Pb, 1 % Ni, and 3 % Sn, in addition to 25 % of organic materials, basically phenolic resins. The precious metals were the main focus of the studies that were carried out in the 1990s. These metals are found in low concentrations, around 250 ppm Au, 1,000 ppm Ag and 100 ppm Pd. In addition to these metals, others can also be found, such as As, Sb, Ba, Br and Bi [6].

Pyrometallurgical processing is done at plants with equipment that was initially developed for the primary metallurgy of copper and lead. Converters of the Isasmelt [7] or Kaldo type are generally used, or even electric ovens for the main scrap processing stage.

PCB scrap, together with non-ferrous metal scrap and other scrap containing precious metals, such as catalysts, are processed at temperatures of around 1,200 °C. Coke is also added, along with the scrap and the by-products. Oxygen or enriched air, in addition to fuel, is injected into the oven. The oxygen or enriched air is blown on top of the oven with a blowpipe, producing an oxidation effect of impurities and agitating the bath and the slag in a similar way as in steel converters. What takes place inside the converter is the melting of the load, the formation of slag, pyrolysis and combustion of the polymers and the selective oxidation of impurities. The coke's function is essentially to provide heat for the initial melting of the load. After the load has melted and the coal and polymers have burnt away, the system's energy is essentially provided by the exothermic oxidation reactions of the metals.

At this temperature, the polymeric materials undergo decomposition and combustion, supplying energy and forming gases. The produced gases transport the volatile metals (for example Be, Hg and Cd) as well as the flame-retardants. Particulate material is also transported by the gases in the oven. In this type of reactor, most of the energy from the combustion processes is lost in the escaping gases. The ceramic materials present in the scrap, and the waste from the by-products that are added, end up in the slag.

During the melting process, oxygen is supplied mainly in order to promote the oxidation of the sulfur, lead, aluminium and iron, which are the main impurities of copper present in the load. The oxidized metals will combine together with the ceramic materials to form the slag.

From a technological point of view, the main concern is not the metallurgical process itself, which is very simple, but rather the cleaning of the off-gas cleaning system, given that the load contains a significant amount of polymers with flame retardants and phenolic resins.

In addition to the dust, which is an inherent part of the metallurgical process, the gas cleaning system, which is connected either to the oven or to the converter, should prevent the formation of VOCs, dioxins and acidic gases. Generally, combined systems are used containing washers, adiabatic coolers and scrubbers [8, 9]. The formation of dioxins during the metallurgical process can be avoided by maintaining such conditions that the gases spend the maximum possible length of time in the oven's hot chamber, or otherwise by promoting an after-burn. In addition to this, there should be a rapid cooling of the gases (quenching) to temperatures below 180 °C [9].

The Umicore process is the one best described by the literature [10]. Umicore recovers 17 metals present in WEEE. Figure 8.1 shows a flowsheet of Umicore's integrated metal smelter and refinery (Au, Ag, Pd, Pt, Rh, Ir, Ru, Cu, Pb, Ni, Sn, Bi, In, Se, Te, Sb, As).

The process is carried out at a temperature of 1,200 °C in an Isasmelt converter. The thermal support is provided by a submerged blowpipe that injects a mixture of enriched air and fuel. Coke is added as both supplementary fuel and as a reduction agent [3].

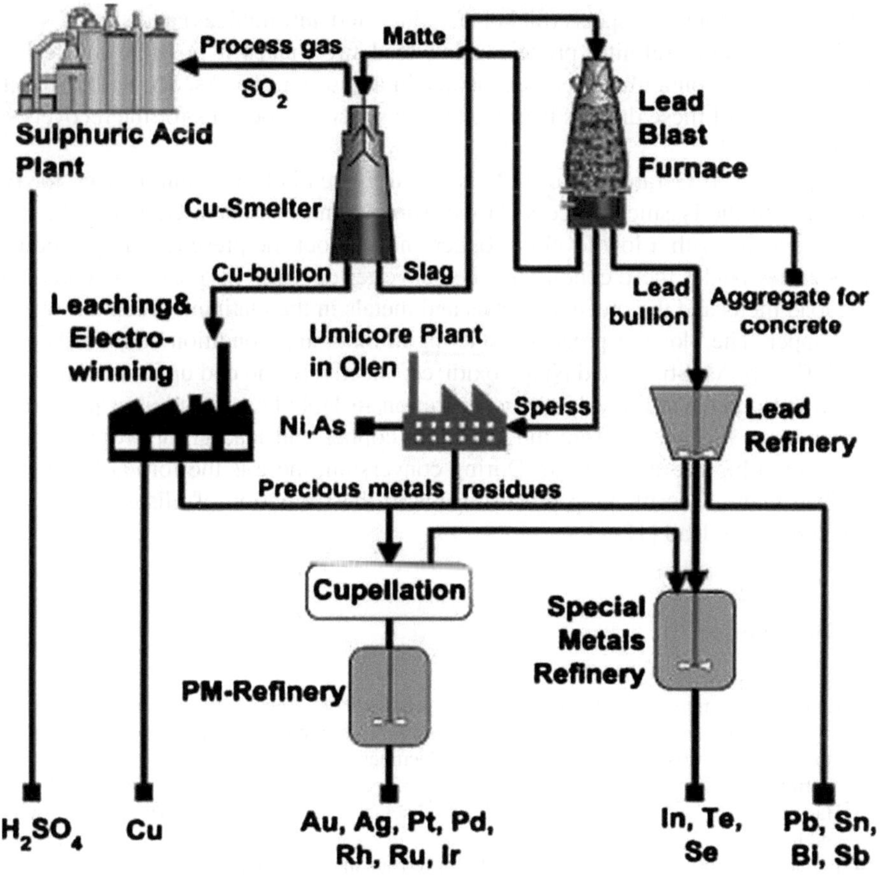

Fig. 8.1 Flowsheet for Umicore's integrated metals smelter and refinery [13]

The blowing process ensures that the aluminum, iron and lead are oxidized and end up joining the slag. As in the primary production process, the precious metals are dissolved in the liquid metallic phase. There are essentially two products, a copper based metallic alloy and a slag.

The metal basically consists of copper and it follows a hydrometallurgical and electrometallurgical route. The metal is therefore leached in sulfuric acid in oxidizing conditions. The copper is converted into aqueous copper sulfate, while the precious metals remain insoluble. The copper sulfate solution then proceeds to electrolytic reduction, producing a metal with a 99.99 % purity level. The acid is recirculated.

The slag, which consists mainly of silicon, calcium, aluminium, iron and lead oxides, is reduced in shaft furnaces, according to the traditional technology used for lead production. In addition to these elements, the slag also contains Bi, Sn, Ni, In, Se, Sb, As and Cu. As in the primary metallurgy of lead, precious metals are also present, mainly silver.

The reduction process that is carried out in shaft furnaces produces bullion lead (95 %), copper mattes, speiss (nickel arsenides and antimonides) and slag.

The traditional refining process using lead fire makes it possible to produce dross or crusts containing precious metals, in addition to Bi, As, Sb, Sn, In, Te and Se. Each one of these drosses is treated using specific processes for the recovery of the metals [11].

According to Gerardo et al. [12], there are essentially two alternatives for the operation of the Isasmelt process. At the Aurubis unit in Lunen, Germany, the load contains scrap with a low level of copper and without the presence of by-products with sulfur. Under these conditions, the converter operates in reduction conditions. The load melts and the precious metals and metals in the platinum group dissolve in the copper. The blowing process results in an oxidizing condition and metals such as Al, Fe, Sb, As, Sb, Pb and Ni are oxidized selectively and end up in the slag.

On the other hand, the Umicore operation in Hoboken in Belgium uses loads containing scrap metals from the platinum, copper and lead group, resulting in a load with a high level of sulfur. During conversion, there is the formation of copper-lead matte. The blowing process leads to the formation of blister copper and slag containing lead [12].

References

1. European Commission Enterprise and Industry (2014) Report on critical raw materials for the EU report of the ad hoc working group on defining critical raw materials. http://ec.europa.eu/enterprise/policies/raw-materials/documents/index_en.htm
2. Worrell E, Reuter MA (eds) (2014) Handbook of recycling. Elsevier, Amsterdam
3. Hagelüken C (2006) Recycling of electronic scrap at Umicore's integrated metals smelter and refinery. Erzmetall 59:152–161
4. Lehner T (1998) Integrated recycling of non-ferrous metals at Boliden Ltd. Ronnskar smelter. In: Proceedings of the IEEE international symposium on electronics and the environment, pp 42–47

5. Veldhuizen H, Sippel B (1994) Mining discarded electronics. Ind Environ 17(3):7
6. Hagelüken C (2006) Improving metal returns and eco-efficiency in electronics recycling. In: Proceedings of the 2006 IEEE international symposium on electronics and the environment. IEEE, New York, pp 218–223
7. Edwards JS, Alvear GRF (2014) Converting using Isasmelt^{TM} technology. http://www.isasmelt.com. Accessed 12 June 2014
8. Kulkarni PS, Crespo JG, Alfonso CAM (2008) Dioxins sources and current remediation techniques—a review. Environ Int 34(1):139–153
9. McKay G (2002) Dioxin characterization, formation and minimization during municipal solid waste (MSW) incineration: review. Chem Eng J 86(3):343–368
10. Hagelüken C (2006) Recycling of electronic scrap at Umicore precious metals refining. Acta Metall Slov 12:111–120
11. Gill CB (1988) Nonferrous extractive metallurgy. Krieger Publishing Co, Malabar
12. Gerardo RF, Alvear F, Nikolic S (2013) Isasmelt^{TM} for recycling of valuable elements contributing to a more sustainable society. In: TMS (The Minerals, Metals and Materials Society). Rewas 2013 100–109
13. Tuncuk A, Stazi V, Akcil A, Yazici EY, Deveci H (2012) Aqueous metal recovery techniques from e-scrap: hydrometallurgy in recycling. Miner Eng 25(1):28–37

Chapter 9
Electronic Waste Recycling

Angela Cristina Kasper, Adjanara Preis Gabriel, Erich Lopes Braitback de Oliveira, Nichele Cristina de Freitas Juchneski and Hugo Marcelo Veit

E-waste contains a variety of valuable materials, such as metals, glass, plastics and other materials (shown in Fig. 9.1). Although recycling processes are usually planned for the metal fractions because of the economical value of this fraction, non-metallic fractions are gaining more attention, since the disposal of such material in landfills, or the burning of plastics can be associated to high environmental contamination [1, 2]. This chapter will discuss the general recycling of material from e-wastes.

9.1 Materials Recycling Considerations

9.1.1 Metals

Nowadays, research and industrial processes are being applied to recover and reuse the valuable materials contained in E-waste, with most efforts focussed on recycling the metallic fractions because of the economic benefits [1].

The recent global increase in demand for metals has made the supply of these metals be perceived as a critical issue. On the other hand, thousands of tons of waste from electric and electronic equipment ais generated each year in the world. As such, end of life electrical and electronics equipment (EEE) is being regarded in recent years as a secondary source of metals.

A.C. Kasper · A.P. Gabriel · E.L.B. de Oliveira · N.C. de Freitas Juchneski · H.M. Veit (✉)
Programa de Pós-Graduação em Engenharia de Minas, Metalúrgica e de Materiais - PPGE3M, Universidade Federal do Rio Grande do Sul - UFRGS, Porto Alegre, RS, Brazil
e-mail: hugo.veit@ufrgs.br

H.M. Veit and A. Moura Bernardes (eds.), *Electronic Waste*, Topics in Mining, Metallurgy and Materials Engineering, DOI 10.1007/978-3-319-15714-6_9

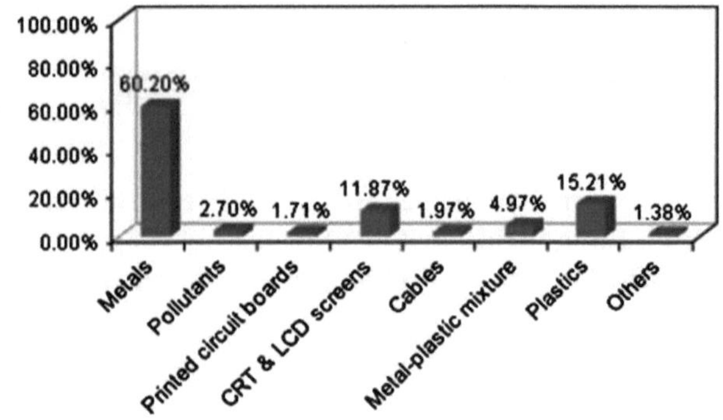

Fig. 9.1 Typical material fractions in WEEE [1]

The large-sized pieces of equipment, such as refrigerators, washing machines, air conditioners and CRT TVs, may be considered as the most important sources of base metals (Al, Cu, Fe, Pb, Sn and Zn). In addition, small devices, such as personal computers, mobile phones, video game consoles and digital devices are being considered as important sources of precious metals (Ag, Au and Pd), while mid-size information and communication technology (ICT) equipment (e.g., printers and fax machines) are an important source of a variety of less common metals (Ba, Bi, Co, Ga, Sr and Ta) [3].

In terms of recycling, the waste from electrical and electronic equipments can be divided into three broad categories: (I) cooling and freezing appliances (C&F); (II) information and communication technology (ICT) equipment; and (III) monitors and televisions (TVs). Each equipment category goes through a different material flow, in which a particular set of technologies to recover the metals can be used [4].

The chain to recycle metals from electronic scrap can be divided into three subsequent steps: collecting, dismantling or pre-processing or mechanical processing, and the final processing. The grinding and the separation techniques (by particle size, by gravity, by magnetic property and by electrical property) are considered to be pre-processing or mechanical processing steps. In the final processing, pyrometallurgical, hydrometallurgical and electrometallurgical techniques can be used.

In pyrometallurgy, high temperatures are used to chemically convert the feed materials and to separate metals and impurities into different phases so valuable metals can be recovered. In hydrometallurgy, strong acidic or caustic aqueous solutions are used to selectively dissolve or precipitate metals, which can later be recovered by precipitation, cementing, and solvent extraction, among others. In electrometallurgy, electrical currents are used to recover metals, e.g., electrowinning of copper and zinc [4].

Typical processes for the recycling of metals from large-sized appliances (refrigerators, fridges, washing machines, air and others) include manual disassembly, grinding, magnetic separation, electrostatic separation (by eddy current) and separation by density. The manual disassembly decouples materials such as wires, circuit boards, coolants, oils from refrigerators and compressors and sends them separately to processing. Magnetic separation is used for the recovery of ferrous metals. Electrostatic separation is used to separate the polyurethane foams and plastics from the metals. Separation by gravity allows the separate recovery of copper and aluminum [5].

In general, the manual disassembly of electronic equipment is performed to remove parts, which may contain metals of economic interest, or to remove toxic substances. The removal of parts that contain valuable materials, such as printed circuit boards, prevents the loss of metals of economic interest (mainly precious metals) during the processing. Moreover, the removal of special hazardous waste, such as mercury from lamps, prevents environmental contamination, allowing a cleaner and safer recycling process. The process of separation and recovery of metals must be adapted to the product and to the materials contained in this product, avoiding unnecessary steps. In the recycling of mobile phones, for example, pre-treatment steps must be restricted to avoid the risk of losing precious metals during processing [6].

The main objective of the recycling of printed circuit boards is the recovery of metals of high economic interest, i.e., copper and precious metals (gold, silver and palladium). Typically, the recycling of the circuit boards is done through pyrometallurgical processes, carried out in copper smelters. However, in recent years an alternative method that combines mechanical pre-treatment and a hydrometallurgical process has been employed. The use of integrated smelters allows a recovery of more than 95 % of the metals of economic interest, especially copper and precious metals [6, 7].

In the processing of lithium-ion batteries, a number of mechanical processes are used, including grinding, melting, atomization and refining, enabling the recovery of cobalt and copper [8].

Currently, units using a combination of pyrometallurgical, hydrometallurgical and electrochemical techniques are able to recover precious metals (gold and silver), platinum group metals (palladium, platinum, rhodium, iridium, ruthenium), special metals (selenium, tellurium, indium), secondary metals (antimony, tin, arsenic, bismuth) and base metals (copper, lead, nickel) [9, 10].

One of the main concerns of the present recycling process is related to the uncontrolled release of toxic chemicals (e.g., lead and PBDE) into the environment. Potential risks to the health of recycling industry workers, resulting from the release of fumes, dust and harmful chemicals into the environment, should also be considered. In this context, the informal recycling processes carried out by collectors with the purpose of recovering metals of high economic value, pollute the environment by the release of toxic substances into the environment [11].

9.1.2 Polymers

The polymers have gained a prominent role, since they are widely employed for various housings of electronic equipment, representing a high percentage by volume of waste generated by such equipment. Polymers in electrical and electronic equipment are highly visible, but there are also many other polymer components hidden inside the equipment, which also provide the necessary structure to the internal parts of the electronics [12].

Polymers are usually chosen for use in electronic equipment frames because of their excellent properties and low costs. Polymers are generally light, have high flexibility, good impact resistance, low processing temperature, low thermal and electrical conductivity, and their properties may be improved by additives (such as fillers, fibers, carbon black, plasticizers, among others).

The polymeric frames of electronic equipment are manufactured in various types of polymeric materials. According to Fisher [13], most polymers used in the electrical and electronics industry are polypropylene (PP), high impact polystyrene (HIPS), acrylonitrile-butadiene-styrene (ABS) and polyphenylene oxide/high impact polystyrene (PPO/HIPS), with a growing use of polycarbonate (PC) blends and PC/ABS. For example, the frames of monitors can contain ABS and HIPS in their composition. These polymers also represent a significant proportion (30–25 %, respectively) of the fraction of polymers in electrical and electronic equipment of all kinds [14]. In mobile phones frames, for example, ABS + HIPS and ABS + PC are found, while in printers HIPS frames are found.

These polymeric frames can also contain flame-retardants, usually brominated. These materials can seriously pollute the environment if not properly discarded. These products, when deposited in inappropriate places, can contaminate groundwater and, if burnt, can pollute the air, which can cause serious harm to those exposed to these substances.

The polymer recycling processes are usually grouped in three general processes: energy recovery, mechanical recycling and chemical recycling [15].

In the specific case of electronic equipment housings, which are normally manufactured or may contain better quality and higher cost polymers, such as ABS, HIPS and PC, energy recycling is discarded at first because the recycled material may present good properties to be reused by industry [16]. Nevertheless, burning is used in some countries as a means of minimizing the landfilling and associated environmental problems of WEEE disposal. However, the firing of the polymers in this type of waste without proper emissions control, can generate furans and polycyclic aromatic hydrocarbons (PAHs) that are extremely harmful to the environment. Some environmental groups have therefore pressured electronics manufacturers to find alternatives for environmentally harmful materials, such as the substitution of the use of brominated compounds as flame retardants in polymers [17]. Péres et al. [18] studied the relationship of cancer mortality in towns near incinerators and their result was surprising. There is a significant increase in the number of cancer cases in cities located near incinerators.

Mechanical recycling of polymers has been identified as the preferred route for the management of electronic waste. In this case, one of the most important factors is to determine the effects of the degradation processes that occurred during the previous life of the polymeric material and during the mechanical recycling. This information is important to predict the shelf life of products made from recycled materials [19].

Generally, there is a degradation of the mechanical properties of polymeric materials and, depending on the structure, a polymer may be vulnerable to environmental effects such as heat, oxygen and its active forms, moisture, air pollutants and ultraviolet radiation. This degradation may involve changes in chemical structure and physical properties [20]. Santos el al. [21] stated that polymeric post-consumer waste will have suffered some kind of deterioration in their physical and chemical properties before recycling. These degradation reactions are usually a result from polymerization, processing, use and the disposal of the polymer.

Even with this potential prior degradation, mechanical recycling is still the most widely used process. It consists in the use of post-consumer polymers that are subjected to procedures that may include some or all of the following steps: grinding, washing, drying, regrinding (flakes) or agglutination and reprocessing. This sequence of processes can lead to degradation of the material, for example, reducing its molecular weight, since the polymer chains are subjected to high temperatures, mechanical stresses, and light, among others [22].

The degradation by mechanical processing has proven to cause changes in the rheology and mechanical properties of polymers such as ABS and HIPS. These changes may be due to the breakup of the PS phase chain and changes in the physical structure of the rubber phase [23]. These polymers undergo physical and chemical changes during their reprocessing and in their ability to interact with impurities, which can alter their composition. This change in composition can change the mechanical properties of recycled polymers [24].

In addition to mechanical and energy recycling, there is also recycling involving chemical processes, such as pyrolysis, hydrolysis, and gasification, among others, to retrieve the polymeric resins from waste. These are processes in which materials are recycled via a chemical process, generating raw materials that can be directly reused or indirectly fed into the production process, i.e., they can be used directly in the production of a new product and then serve as an input in the process [15]. Although this type of recycling allows for the obtaining of basic raw materials such as monomers, it is still not widely used, mainly because of its complexity and high cost.

Gabriel et al. [25] studied the recycling of cathode ray tubes (CRT) monitor frames. This study compared the same brands with different colors and years of manufacture. The results show that the housings have the same composition, with no difference when the different years of manufacture are compared. This comparison is important because when the composition of the material to be recycled is known, the process is easier and does not require a prior separation. Kasper et al. [26, 27] studied the recycling of mobile phones by testing different brands and models of mobile phones. From the analysis, they were able to observe that the housings have the same composition and are therefore suited for recycling. Mechanical testing was also performed, and the blends were compared to the virgin material. This proved the potential use of recycled material in the production of new cell substrates, or of any other part /component that requires similar characteristics.

Taurino et al. [28] studied the recycling of televisions and computer screens, and pointed out that, before proposing a recycling process, a complete characterization of each WEEE housing is necessary in order to obtain all information about differences in the physical and chemical properties of each device. According to the authors, mechanical recycling, especially the preparation of blends of recycled polymers, can be an interesting possibility to recycle the plastic waste from computers and televisions.

Recently, cathode ray tube monitors have become outdated and new technologies such as LCD and LED are taking over. LCD monitors are lighter, thinner and more energy-efficient, they do not emit radiation (in contrast to CRTs), and have a longer life cycle. However, LCD screens are also classified as hazardous waste. This means that if these are not screens recycled, then they should be disposed of in appropriate locations for hazardous industrial waste, without contact with the environment [29]. An advantage of replacing CRTs by LCDs is that it would significantly reduce the concentration of lead in electronic waste [30].

9.1.3 Ceramics

The development of electronics owes a lot to the use of ceramics. This class of materials is the foundation of modern electronics equipment. The use of silicon enabled the creation of the microchips that are found in every piece of electronic equipment.

The properties of ceramics, such as their insulating, semi-conducting, superconducting, piezoelectric and magnetic properties, have turned them into a critical material for microchips and others components found in cell phones, computers, televisions, and other consumer electronics [31].

A ceramic material is any inorganic and non-metallic substance. It is usually formed by a metallic and a non-metallic element, such as SiO_2 or Al_2O_3. Ceramic materials have a wide range of physical properties that make their use possible in many products [32].

The history of modern electronics began in 1954, when Gordon Teal and his team announced the invention of silicon transistors. Until 1954, transistors were made of germanium, an expensive element, and showed little potential for substituting vacuum valves. Gordon's discovery, however, enabled the use of silicon in integrated circuits commonly known as microchips [33]. The invention of silicon microchips let to the creation of microprocessors and revolutionized the world.

Since the discovery of microchips, their manufacture has evolved to a microscopic scale, and, today, to a nano scale [34]. As a result of this development, the amount of ceramic materials used in the manufacture of microchips is decreasing with time.

Ceramic materials can be found in a variety of components in electrical electronic equipment. Table 9.1 present the main components that have any ceramic material in their composition.

Table 9.1 Components with ceramic materials

Component	Element of the component	Compound/element
PCB	Woven fiberglass cloth	Silica
Monitors	Glass	Silica
Ceramic capacitors	Core, casing	Ba, Ti, Mg, Nb, Zn, Ta oxides
Microchips	Core, casing	Silicon
Processors	Core	Silicon
Ceramic antennas	Casing	Silica and other oxides
Cables	Lump/ferrite bead	Iron oxides
Batteries	Electrolyte	Ceramic oxides

Most electrical electronic equipment has a printed circuit board (PCB) with one or more microchips for control. PCBs and CRT monitors are the two main sources of ceramic materials in WEEE. CRT glass is an important source of ceramics and its recycling is discussed in a separate chapter.

Woven fiberglass cloths are by far the most common material used in the manufacture of PCBs [35]. The two most common types of PCBs are FR-4 and FR-2. These PCBs are made of fiberglass reinforced with an epoxy resin and cellulose paper reinforced with a phenolic resin, respectively [36]. The composition of PCBs varies according to the type. Silicon, aluminum, calcium, boron, sodium, potassium, magnesium, iron and titanium oxides are usually employed in the manufacture of PCB fiberglass. These oxides should therefore be expected in the recycling process.

The recycling of ceramic materials basically occurs through a physical separation process or a chemical process. Most ceramic materials are hand picked in the disassembly process. The recyclers usually employ this process to facilitate segregation and to increase the economic value of recycling.

Ceramic materials can be separated through a physical separation process. This process includes manual disassembly, magnetic separation and eddy current separation. Many recycling companies utilize this separation process prior to a smelting or a hydrometallurgical process. The physical separation process is a mechanical process for the separation of the magnetic, non-ferrous and non metallic materials of WEEE.

The process begins with the disassembly of the components from the electro-electronic equipment. Usually, this is a manual process and it is performed to facilitate the comminution process. The comminution process is employed to reduce the particle size so as to improve the separation of materials in the next process. Generally, shredder mills are used and the final particle size is <2 mm. The particle size plays an essential role in magnetic and eddy current separation, since the WEEE components have different chemical compositions and are difficult to separate.

The magnetic separation step separates the comminuted material into two fractions: the magnetic fraction and non-magnetic fraction. The magnetic fraction contains magnetic materials such as iron and nickel and the non-magnetic fraction contains the non-ferrous metals and non-metallic materials. Since ceramic materials are not magnetic, they remain in the non-magnetic fraction along with other non-ferrous metals and polymeric materials. This separation is performed on belts in a crossbelt system line.

The non-ferrous fraction from the magnetic separation step is separated in the eddy current separator. This equipment divides the non-metallic fraction (NMF) from the non-ferrous fraction. Ceramic materials and polymeric materials are part of the NMF, and metals like aluminum, copper, silver, zinc, gold, tin and lead remain in the non-ferrous fraction [37].

The resulting non-metallic fraction is composed of fiberglass, polymeric resin, and minor amounts of metals like copper and solder alloy [38, 39]. The exact composition of this fraction varies according to the source of WEEE.

In some companies, polymeric materials are separated from ceramic materials by using the difference in densities between these materials. The polymers are then used as fuel for energy generation in incinerators or as polymeric raw material. The remaining fraction consist basically of ceramic materials that are usually landfilled [40].

The destination of the non-metallic fraction depends on its chemical composition. If the amount of polymers present in the non-metallic fraction favors the criterion of calorific power of the incinerator, the fraction is employed in energy generation. However, if the amount of ceramic materials is not viable for incineration, the fraction is landfilled.

Smelting or incineration of the non-metallic fraction always results in the formation of ashes. Most ceramic materials are unavailable for energy generation because of their low calorific power and high silica content. However, many ceramic materials are smelted regardless of previous physical separations processes because they are difficult to remove in the hand picking step, or because they have no economic value. The ceramic materials remain in the slag fraction of the smelter [9].

Ceramic materials are also found in batteries. Lithium-ion and Ni-MH batteries have many types of ceramic oxides in their composition [41]. The recycling process of these batteries usually involves smelting and hydrometallurgical processes and generates large amounts of calcium, aluminum and silicon oxides.

The Umicore Company developed a recycling process for batteries using smelting and hydrometallurgical techniques [42]. The process consists of smelting the batteries to form a slag and a metal alloy. The metal alloy contains the metals of interest, such as nickel and cobalt, which are further treated for recovery. The slag is composed mainly of calcium, aluminum and silicon oxides with minor amounts of iron, manganese and lithium oxides.

The slag is treated in a hydrometallurgical process for the separation of lithium. Silicon, aluminum, calcium and magnesium are by-products of this process and they are sold to the construction industry as an aggregate material, raw material for concrete or as filling material in the manufacture of concrete.

Commercially, the most widely chosen destination for ceramic materials, especially NMFs, is landfilling. This occurs because the majority of ceramic materials have little economical value compared to other materials, such as metals and even some polymers.

Many researchers have proposed alternative destinations for these materials. These alternatives focus on the recycling of ceramic materials for the manufacture of new ceramic products or for the manufacture of ceramic-based materials.

Ceramic materials like ferrites and isolating ceramic components are recycled for the manufacture of new ceramic materials. These components are removed in the hand picking step and their recycling has been well established.

The recycling of ceramic materials for the manufacture of ceramic-based materials is not widespread in the recycling industries. The main recycling alternative for ceramic materials is as an aggregate material in the manufacture of cement for construction. Niu and Li [43] investigated the use of PCBs in the cement solidification process for ordinary landfilling. Although it is not a direct recycling process of the NMF, the study shows a recycling alternative for this material.

Another recycling alternative of NMFs is using them as fillers for thermosetting resin composites and thermoplastic resin composites. Guo [44] studied the use of NMFs as a replacement of wood flour in the manufacture of phenolic molding compound, a thermosetting resin matrix composite. In another study [45], Guo proposed the use of NMFs as a filler in an unsaturated polyester resins for the manufacture of sewer grates, park benches, fences and trays. The recycling of NMFs in thermoplastic resin composites was studied by Zheng [46], the method consists of mixing polypropylene with a modified NMF to generate a NMF/PP pellet composite.

In relation to glass cullet, there are some alternatives to use this waste [47]. Research has been conducted into the use of cullet in conventional clay-based fired ceramics [48]. Another study has examined the use of glas as a replacement for feldspars in porcelain stoneware tiles [49]. Glass is a hard material, which makes it suitable for use as abrasive medium. Benjarungroj et al. [50] obtained good results with media made from molded and sintered glass powder used for vibratory mass finishing instead of polyester resin-bonded ceramic particles.

9.2 Printed Circuit Boards

The printed circuit board (PCB) is one of the basic components of electrical and electronic equipment, and it is commonly used in the telecommunications, information technology, automation, entertainment equipment, and automotive industries, among others.

A printed circuit board basically consist of a non-conductive substrate, the printed conductive circuits and electrical components. In some cases, the term ***printed circuit board*** (PCB) is used as the definition of a non-mounted board, in other words, the board containing only the substrate and the printed conductor circuit. In this book, we use the term ***printed circuit board*** for the mounted boards, in other words, boards that also contain the electronic components.

PCBs are composed of various types of metallic and nonmetallic materials. The average weight of these materials in a PCB is approximately 30 % of polymers, 30 % of refractory oxides (mainly silica, alumina, rare earth oxides), and 40 % of metals (copper, iron, tin, nickel, lead, aluminum, gold, silver, among others) [51, 52].

PCBs do not have a standard design, in other words, each board is manufactured according to the operating needs of the electronic system of the intended product. They can be classified by their technology: single-sided (with conductive circuits on only one side of the substrate), double-sided (with conductive circuits on both sides of the substrate) and multilayer (with circuits between the substrate layers, which may vary from 4 to 16 layer) [53].

Another way of classifying PCBs is based on the material from which its substrate is made. In this classification, the boards are termed commercially as FR-1, FR-2, FR-4 (where, FR indicates the presence of the flame-retardants) and CEM-x.

The boards of type FR-1 and FR-2 are made of cellulose paper impregnated with phenolic resin, while the type FR-4 boards are made of fiberglass and epoxy resin. The boards of the type CEM-x are made of composites produced with fiberglass and cellulose paper with epoxy resin. Boards can also be made of Polytetrafluoroethylene (PTFE) and polyester. In computers, mobiles phones and other communication equipment, FR-4 boards are usually used, while in televisions and home appliances FR-2 boards are used [54].

The conductive circuits used for most applications are made of copper, although in some specific cases aluminum, nickel, chrome or other metals may be used [55]. The electrical components include chips, semiconductors, connectors, capacitors, and others. These components may contain metals such as iron, tin, nickel, lead, aluminum and precious metals (gold, silver, palladium, etc.). In addition, other elements, such as: tantalum (Ta) can be used in the capacitors; gallium (Ga), indium (In), titanium (Ti), silicon (Si), germanium (Ge), arsenic (As), selenium (Se) and tellurium (Te) can be found in the chips and other units; tin (Sn) and cadmium (Cd) can be used in solders; and gallium (Ga), silicon (Si), selenium (Se), germanium (Ge) can be applied in semiconductors [52].

The incorrect disposal or the incineration of PCBs without an emissions control system can cause serious environmental problems, particularly through the pollution by mercury (Hg), cadmium (Cd), lead (Pb), arsenic (As), dioxins and furans. These substances end up polluting the air, soil or water. Many of these pollutants can persist in the environment, since they bioaccumulate up the food chain and pose a risk to human health and to the environment [56–60].

In Europe, electrical and electronics equipment are subjected to the directive on the restriction of certain hazardous substances (RoHS), and to the directive on collection, treatment, recycling and disposal of waste from electrical and electronic equipment (WEEE). The WEEE Directive gives manufacturers the responsibility for the collection, recycling and/or disposal of waste from electrical and electronic equipment. The RoHS directive also provides for the restriction of the use of six hazardous substances in the manufacture of electrical and electronic products. According to the RoHS, the percentage of cadmium should not exceed 0.01 % (m/m) in homogenous materials of electric and electronic equipment and their waste, while the percentage of polybrominated biphenyl (PBB), polybrominated diphenyl ethers (PBDE), lead, mercury and hexavalent chromium should not exceed 0.1 % (m/m). Countries like China, Japan, Brazil, and others have their own policy regarding the issue of waste from electrical and electronic equipment (WEEE) [61–63].

To reduce harmful emissions and to control environmental pollution, the electronics industry has used tools such as eco-design, life cycle analysis (LCA) and end of life (EOL) analysis in recent years to improve its environmental performance. These tools make the assessment of the environmental impacts and economic losses associated with their production, consumption and disposal possible. They also enable the design of actions to increase the lifespan of equipment and/or strategies to reuse and recycle their components [64, 65].

Estimates indicate that each year about 40 million tons of waste from electrical and electronic equipment is generated globally. Circuit boards constitute between 3 and 6 % of the total weight of e-waste [66, 67]. In Germany, for example, 10,300 ton of waste from printed circuit boards was accumulated in 2009 [68].

Generally, the concentration of precious metals in the printed circuit board is much greater than the concentration of precious metal in ores. Currently, the ores from which gold and palladium are extracted, for example, contain less than 10 g/t of these metals, while the printed circuit boards of personal computers may contain over 250 g/t gold and 110 g/t palladium. Besides this obvious economic advantage, one should take into account that the environmental impacts associated with the secondary production of precious metals are much lower than those associated with the extraction of these metals through mining [69]. Obviously, the quantities of precious metals present in printed circuit boards vary greatly depending on the equipment and on the functions for which they were intended.

Typically, a PCB from a PC contains 7 % Fe, 5 % Al, 20 % Cu, 1.5 % Pb, 1 % Ni, 3 % Sn and 25 % organic compounds, together with (in parts per million - ppm) 250 Au, 1,000 Ag and 100 Pd. In addition, traces of As, Sb, Ba, Br and Bi are present [70]. A PCB of a mobile phone may contain on average 20–30 % copper, 6–8 % ferrous metals, about 5 % nickel, 2.5 % tin, 1.1 % zinc; 250 mg of silver, 24 mg of gold, 9 mg of palladium, and other metals. The percentage of materials depends on the model and year of manufacture [52, 71]. So even if the recycling of printed circuit boards is quite complex, especially due to the variety of materials found in it, the presence of precious metal or of metals of high economic value, such as copper, silver and gold, among others, makes PCBs a quite interesting raw material to recycle.

The main challenge for the recycling of printed circuit boards is the fact that more than 60 % (by weight) of it is composed of non-metallic materials. The non-metallic materials of PCBs mainly consist of thermosetting resins and glass fibers. Thermosetting resins cannot be remelted or reformed because of their network structure. Incineration is not the best method for treating nonmetallic materials because of inorganic fillers such as glass fiber, which significantly reduces the fuel efficiency. Currently, disposal in landfills is the main method for treating the non-metallic materials in PCBs. However this action can cause secondary pollution and waste of resources. Some studies have suggested the use of a percentage of this material (about 20 %) together with resins in the manufacture of other types of plates or, in lower percentages, as filler in the manufacture of other plastic materials [27, 72].

The informal recycling of printed circuit boards, where the scraps are incinerated to obtain precious metals without controlling emissions, or where acids are used for leaching without controlling effluents, are considered causes of serious problems to the environment and people's health. Studies show that regions where informal recycling plays an important economic role have serious environmental and health problems due to high concentrations of lead (Pb), polybrominated (PBDEs) and polychlorinated dioxins and furans, in addition to polybrominated dioxins and furans (PCDD/Fs and PBDD/Fs). The main problems related to this type of pollution are endocrine disruption and neurotoxicity that can persist over generations [58–60].

It is therefore essential that the recycling of printed circuit boards is carried out correctly, avoiding contamination of the environment.

In the recycling of printed circuit boards, a single process or a combination of two of the following processes may be used: mechanical processing, hydrometallurgical processing, pyrometallurgical processing and bioleaching.

9.2.1 Mechanical Processing

In primary metallurgy, mechanical processing has been used as one of the treatment and processing steps of metals and minerals. In the recycling of printed circuit boards, mechanical processing has been generally used as a pretreatment for the recycling of the metallic fraction [73].

Operations that are part of the mechanical processing are: disassembling, grinding, classification and separation of materials. The process of separation of materials can be carried out through differences in density, weight, particle size, magnetic properties or electrical properties.

The aim of grinding is to release metals from the base plates of the PCB waste. Grinding is very important because it provides the appropriate characteristics to materials for the subsequent separation steps. If grinding is not well done, such problems as insufficient decoupling of metals and non-metals, excessively sprayed particles, incomplete grinding, material plugging, and equipment fever may arise [74]. Generally, the grinding is performed in two steps. In a first step the boards are reduced to a size <10 mm and then the particles are reduced to sizes that permit the release of metals in a second step (<2 mm). The particle size should not be less than 0.3 mm because the degree of separation between metals and non-metals is reduced when the particles are very small [75, 76].

After grinding, it is important to make a particle size separation prior to use other separation processes. This separation based on the size of the particles is necessary because the particle size influences other material separation processes directly (density, magnetic or electrostatic separation).

Methods such as gravimetric (or concentration) separation and dense medium separation can be used to separate the metallic fraction from the non-metallic fraction of PCBs. When applied to materials with particle size exceeding 0.6 mm, separation by gravity using zig-zag type classifiers may lead to a concentration of up to 95 % of the metal components in the heavy fractions. However, when the particle size is less than 0.6 mm, this concentration of metallic components is reduced to only 60 %. Another factor to be taken into account when using this method for separating or concentrating metal, is the fact that some types of metals, perhaps because of their density, have a recovery rate under 60 %, which demonstrates that separation by gravity is not a good alternative for some metallic elements [77].

A zig-zag type concentrator was used in a study that combines vacuum pyrolysis with mechanical processing steps (grinding, particle size separation and gravity separation) in the recovery of metals of disassembled PCBs. In this case, the pyrolysis was used for the recycling of the organic fraction of PCBs, while the mechanical processing aimed to release copper for later retrieval. The results show a recovery rate of 99 % of the copper in PCBs [66]. Studies using Falcon-type concentrators for the separation of fine particles (1.0 mm) demonstrate that with this type of separator it is possible to concentrate more than 92 % of metals in the heavy fraction, which would allow a recovery rate of over 97 % [78].

The concentration of metals using the dense media separation technique was tested with relative success by Veit et al. [79]. In this work, a mixture containing Tetrabromoethane (TBE) and 27 % acetone (2.5 g/cm^3 density cutoff) was tested to separate the metallic fraction (heavier fraction) from the non-metallic fraction (lighter floating fraction). However, it should be noted that using dense media separation has the disadvantage of using toxic and expensive organic solutions.

The most widely employed process for the separation and concentration of metals in the recycling of printed circuit boards is one that combines magnetic separation and electrostatic separation. In the first stage (magnetic separation), magnetic metals (iron, nickel and other ferrous metals) are separated or concentrated. In the second stage (electrostatic separation), metals susceptible to electrostatic induction (copper, tin, lead, etc.) are separated from the non-metallic materials.

The choice of the magnetic separator to be used in the processing of PCB waste depends on several considerations, the most important being the particle size distribution and the distribution of the magnetic product [80]. The use of two stages of magnetic separation, one using a field of lower intensity and the other using a field of higher intensity can be an alternative to increase the concentration of magnetic metal in the magnetic fraction. However, the use of very high intensity magnetic fields may very well increase the concentration of magnetic metals, such as iron and nickel, but it can also result in other metals, which are less susceptible to magnetic fields, being attracted and ending up in the magnetic fraction [77].

Electrostatic separation is considered an effective way to separate of concentrate metals and non metals from printed circuit boards, and the Corona-type separators are the most widely used [81–83]. The electrostatic separation technique using the Foucault current (eddy current) was used to recover aluminum scrap from PCBs. With this technology, it was possible to recover more than 90 % of aluminum, with 85 % purity [84, 85].

A combination of mechanical processing steps, comprising of grinding and particle size separation; and of electrostatic separation and magnetic separation, was successfully tested by some authors in the concentration of metals from printed circuit boards. In these studies, a magnetic fraction containing, on average, 45 % of iron; another metal fraction (obtained in the electrostatic separation) containing between 50 and 60 % copper, 24 % tin, 8 % lead; and a non-conductive fraction containing polymers and ceramics was obtained. The use of these methods enables a high percentage recovery of these metals [52, 73, 86, 87].

9.2.2 Hydrometallurgical Processing

Hydrometallurgical processes have certain advantages over Pyrometallurgical processes in the recycling of PCBs, such as lower capital costs, no formation of gas or dust, operational selectivity and suitability for small-scale applications. These advantages make hydrometallurgical processes an alternative for the treatment of PCB waste [88].

The research using hydrometallurgical techniques for the recovery of metals from scrap from electrical and electronic waste is mainly focused on the leaching of metals of economic interest, such as gold, silver, copper, tin, and nickel, among others. The hydrometallurgical processing of PCBs most often involves two stages of leaching. In the first stage the base metals are extracted, with the precious metals being extracted in the second stage .

In the first stage, mineral acids such as HCl, H_2SO_4, HNO_3, $HClO_4$ can be used. Hydrogen peroxide, which is a strong oxidizing agent, is normally used in combination with the acids to increase metal extraction. In the leaching of precious metals, solutions containing aqua regia, cyanide, or more recently, thiourea, thiosulfate or halides can be used.

Before starting the recovery of metals from PCBs through hydrometallurgical processes, a pretreatment is performed in most cases. This pretreatment can include selective removal (dismantling) or treatment involving mechanical processing. In the dismantling, the components and devices can be separated into various fractions (chips, semiconductors, connectors, capacitors, base plate, etc.) and treated separately according to their characteristics. This pretreatment aims to improve process efficiency and to decrease the amount of solvent to be used in the leaching [88].

The use of nitric acid, hydrochloric acid or sulfuric acid in the dissolution of the base metals present in printed circuit boards has been the subject of several studies. Kinoshita et al. [89] used two stages of leaching, with different concentrations of nitric acid, to recover copper and nickel of non-mounted printed circuit boards (PCBs without electrical components). In the first step they used a diluted nitric acid solution (0.1 M), obtaining a rich nickel solution with small quantities of copper. In the second stage, a more concentrated nitric acid solution (1.0 M) was used, obtaining a solution rich in copper with virtually no impurities. The effects of the presence of copper ions (II) in the dissolution of copper with chloride and hydrochloric acid was also studied by other researchers [90].

Studies indicate that the dissolution of tin and lead with nitric and/or hydrochloric acid and subsequent precipitation of tin from H_2SnO_3 (metastannic acid) allowed the recovery of 97 % of the tin from the boards [91]. The use of a solution containing H_2SO_4–$CuSO_4$–NaCl enables the leaching of more than 90 % of copper, iron, nickel and silver, and 58 % of palladium [92].

Studies realized by Havlik et al. [93] showed that the recovery of copper and tin using a pyrolysis step followed by a leaching step can be improved in the case of copper, but deteriorated in the case of tin. The results indicate that the amount of dissolved copper obtained from pyrolyzed samples was greater than the amount obtained from non pyrolyzed ones, while the quantity of dissolved tin obtained from a pyrolysed sample was smaller than the amount obtained from the non pyrolyzed one.

Some authors have used aqua regia as leaching agent of the precious metals (gold, silver, palladium) contained in PCBs, obtaining leaching rates of 98 % for silver, 97 % for gold and 93 % for palladium [94]. However, although the dissolution of gold from PCBs is very effective, this type of processing appears to have little practical application, since during the leach, the aqua regia releases nitrous oxide and chlorine vapors that may contaminate the environment and cause poisoning.

Cyanide has been used as leaching agent for gold and silver by the mining industry for over a century with some advantages, such as: low cost, low reagent consumption and operation under alkaline conditions. Likewise, solutions containing cyanide are used to dissolve the gold and silver present in PCBs. In this case, the gold recovery occurs by reducing the cyanide complex. However, the high degree of toxicity of these compounds, both to workers and to the environment, has led researchers to study leaching processes that do not use cyanide as leaching agent. Among these leaching processes, leaching with thiourea, thiosulfate and halide have been studied [95].

Jing-Ying et al. [96], using solutions containing 24 g/L of thiourea and 0.6 % of Fe^{+3} in the recovery of gold and silver from the PCBs of mobile phones, obtained leaching rates of 90 % for gold and 50 % for silver. Gurung et al. [97], using solutions of thiourea in samples of PCBs previously crushed and calcined (at 750 °C for 6 h), obtained leaching rates for gold and silver exceeding 90 % after 6 h of leaching. The addition of ferric ions as oxidizing agents reduced the leaching time to 2 h.

The high concentrations of copper found in PCBs significantly affects the dissolution of gold. As such, a pretreatment with sulfuric acid and hydrogen peroxide to recover the copper beforehand, enables greater dissolution rates of gold in solutions containing thiosulfate or thiourea, for example. According to Birloaga et al. [98], the use of a preliminary stage for the recovery of copper, followed by the dissolution of gold with thiourea, enables the recovery of more than 90 % of copper and about 70 % of gold.

The commercial application of thiourea as leaching agent for gold and silver is hampered, mainly, by three factors: thiourea is more expensive than cyanide; consumption is higher because thiourea is readily oxidized in solution; the gold recovery step is still undeveloped [95].

The use of ammonium thiosulfate as leaching agent for the gold from the PCBs of electronic scrap was also studied. Oh et al. [99] obtained leaching rates of more than 95 % for gold (in 48 h) and about 100 % for silver (in 24 h) using ammoniacal thiosulfate as leaching agent of the electronic PCB scrap. Likewise, Ha et al. [100] obtained leaching rates for gold of 98 % (in 2 h) for samples of PCBs from the keyboard scrap from mobile phones, and of 90 % (in 10 h) for samples of others types of PCBs from mobile phones, using solutions containing 0.12 M of ammonium thiosulfate, 20 mM of copper II and 0.2 M of ammonia.

Thiosulfate leaching has the advantage of high selectivity, nontoxicity and noncorrosivity. The main problem with thiosulfate leaching is the high consumption of the reagent during extraction [95, 101].

The recovery of metals of interest from leach liquors is an essential step in hydrometallurgical processing. In this recovery, a variety of techniques can be used, such as: electrometallurgy, cementation, solvent extraction, adsorption on activated carbon and ion exchange. The selection and development of metal recovery processes are based essentially on such factors as: the reagent of the leaching system, the concentration of metal(s) and impurities.

One way to recover copper from leach liquors implies the use of electrochemical processes Kasper et al. [52] and Veit et al. [87] used the electrochemical process to recover copper from solutions obtained from samples of metals previously concentrated by mechanical processing steps (grinding, particle size separation, magnetic and electrostatic separation), resulting in recovery rates of copper exceeding 92 %.

Scott et al. [102], studying the recovery of metals from the pickling solutions used in the manufacture of PCBs, analyzed two recovery processes. In the first process, only the electrochemical recovery of metals was used, while the second process consisted of a combination of electrowinning (for copper) and metal precipitation (for tin and lead). The metals were recovered from the precipitate by a pyrometallurgical process.

Cementation is one of the most effective and economical techniques used in the recovery of metals obtained in hydrometallurgical processes, including the recovery of precious metals from leached liquors. The advantages of the cementation process include the relative simplicity, the easy process control and the ability to recover valuable metals.

Cementation with zinc (also known as the "Merrill-Crowe process") is a process that has been known since the late XIX century and it is used to recover gold from gold-cyanide solutions. The main reactions occurring in this process are the deposition of gold in the cathode and the anodic corrosion of zinc [103]. Copper, which is relatively cheaper and has a lower reduction potential than gold and silver, and, at the same time, has a higher reduction potential than other base metals, can also be used in the cementation of gold and silver liquors containing cyanide or thiourea. The results of the cement tests for recovering gold and silver with copper powder showed a complete recovery of silver and a recovery of about 80 % of gold [104, 105].

The reduction-precipitation with metal hydrides has been used to reduce gold and silver from the leach liquor at a commercial scale. The most popular metal hydride reductant is sodium borohydride. The recovery of gold from solutions of thiourea, thiosulfate or thiocyanate by reduction-precipitation using solutions containing 12 % $NaBH_4$ and 40 % NaOH was successfully carried out. The results showed that gold ions can be efficiently reduced to metallic gold from acidic solutions, even at room temperature. This reaction can be performed either in very dilute or more concentrated solutions [106].

Gold and silver can also be recovered by adsorption on polymeric adsorbents, activated carbon or other adsorbents from biomass, such as carbonized rice husks and persimmon tannin gel [107–110].

The recovery of metals from the leaching solution by solvent extraction has been successfully tested by some authors. Fray and Park [94] used extraction with toluene, followed by cementation with dodecanethiol and sodium borohydride to selective recover the gold extracted from PCBs with a aqua regia solution. Kinoshita et al. [89] used the commercial solvent LIX984 in the selective extraction of copper from a solution that contained copper, nitric acid and nickel impurities, obtained from the leaching of PCBs.

In the recovery of gold, some organophosphorus derivatives, guanidine derivatives and mixtures of amines- organophosphorous derivatives products have been studied. The commercial solvent LIX79 can be used in the extraction of gold in alkaline solutions containing cyanide, while Cyanex 921 can be used for all pH ranges [106].

Ion exchange resins can be used in the recovery of gold from leaching solutions. Such commercial resins as Dowex G51, Dowex 21 K and Amberlite IRA-410 have been tested with success in the recovery of gold from thiosulphate solutions. These resins are gel-type with a polystyrene divinylbenzene matrix, with quaternary ammonium in their functional groups [106].

9.2.3 Biohydrometallurgical Processing

The recovery of metals by Bioleaching can be considered one of the most promising technologies tested in the last decades to obtain metals, arousing great interest from mineralogy and metallurgy companies. It can be applied to a variety of metals such as copper, nickel, cobalt, zinc, gold and silver. A series of studies have been conducted in recent years to test the use of such techniques to obtain metals from ores. There are still relatively few studies, however, on the possibility of using this technique for the recovery of metals from printed circuit boards [111, 112].

Biohydrometallurgical processes are essentially hydrometallurgical processes that use microorganisms (i.e., bacteria, archaebacteria and fungi archaea) to improve the dissolution of metals from ores, concentrates or waste. In these processes, the use of microorganisms is based on their characteristics to oxidize or use the organic and inorganic substrates to generate a leaching agent for the dissolution of metals. The microorganisms studied in the recovery of metals are *Acidithiobacillus ferrooxidans, Acidithiobacillus thiooxidans, Leptospirilum ferrooxidans and archaeas.* In some cases, it is possible to use microorganisms such as the cyanogenic bacterium *Chromobacterium violaceum*, which are capable of generating cyanide ions, promoting the dissolution of precious metals in the form of a cyano-complex [106, 113].

In some cases, leaching columns are used. In these cases, the columns are filled with grinded printed circuit boards scraps. The column bed is then irrigated by a sulfuric acid solution with a controlled pH and irrigation rate. The solution percolates the bed of electronic scrap containing sources of nutrients such as nitrogen (N), phosphorus (P) and potassium (K), in addition to bacteria and archaeas. At the base of the column, air is blown in from top to bottom to ensure the supply of oxygen and carbon dioxide in the reaction system. For the bioleaching process to take place, it is essential to guarantee the availability of dissolved oxygen on the surface of fragmented particles and carbon dioxide (CO_2), which is used as a carbon source and is necessary for the synthesis of cellular material used for producing new cells, together with the other above mentioned nutrients [114].

Ilyas et al. [113] studied the possibility of bioleaching metals from printed circuit boards using strains of *acidophilic thermophilic, acidophilic chemolithotrophic* and *heterotrophic* bacteria, including *Sulfobacillus thermosulfidooxidans.* They obtained leaching rates of 81 % for nickel, 89 % for copper, 79 % for aluminum and 83 % for zinc. Zhu et al. [115] used a mixed culture of acidophilus bacteria (MCAB) to extract metals from PCBs. Under optimal test conditions, they obtained 96.8 % of leaching efficiency for copper after 45 h, and 88.2 % and 91.6 % for aluminum and zinc after 98 h, respectively. In this case the optimal conditions were: initial pH 2.00, 12 g/L Fe (II), 12 g/L of metal concentrate, dosing inoculation amount of 10 %, and 60–80 mesh particle size.

The use of micro-organisms in the recovery of metals from electronic waste can be an economic alternative since it's process that requires a lower initial investment and low energy consumption, especially if compared to pyrometallurgical processing, given that the latter requires a high energy consumption and requires a high investment for deployment [115].

9.2.4 Pyrometallurgical Processing

The pyrometallurgical processing is the most traditional technology for the recovery of non-ferrous and precious metals from waste electronics. Some of the pyrometallurgical processes include incineration, smelting in a plasma arc furnace or blast furnace, drossing, sintering, melting and reactions in a gas phase at high temperatures. In these processes, the waste is burned in a furnace or in a melting bath to remove the plastic material. In this phase the oxides of metals and refractory oxides go into a slag phase and the metals of interest can be recovered [116].

The main advantages of pyrometallurgical processes include the acceptance of any kind of electronic scrap, the fact that it does not require a pre-treatment and has few steps. The main disadvantages are the possibility of generating toxic gas emissions, the loss of metal by volatilization, the increase in the amount of slag, due to the presence of components such as glass and ceramics, which causes a higher loss of noble metals and base metals, the low rates of recovery of some metals (e.g., Sn and Pb) or the impossibility of recovering other metals (e.g., Al and Zn) [117].

Incineration is the most common way of getting rid of polymeric materials and other organic materials present in printed circuit boards. The melting of the material produces impure metal alloys, which can then be refined by electrolytic or pyrometallurgical processes. In this case, the thermal energy given off from such materials is not recovered [118]. The incineration of PCBs in a traditional combustor designed for municipal solid waste can be very dangerous. Copper, for example, is a catalyst for the formation of dioxins during the incineration of brominated flame retardants (BFRs) used in most PCBs [106]. Another problem with the incineration of PCBs is the low efficiency of burning because they contain inorganic materials such as glass fiber [45].

Pyrolysis can be considered as an alternative method to recycle PCBs. In the process of pyrolysis (heating without oxygen), the organic material is decomposed forming products of low molecular weight (liquid or gas), which can be used as fuel or chemical feedstock. Vacuum pyrolysis has several advantages over other methods of pyrolysis because of the short residence time of organic vapors in the reactor and the low decomposition temperature, which reduces the occurrence and severity of side reactions [119].

The use of pyrolysis for the recovery of waste from electrical and electronic equipment (WEEE) has been studied by some authors. Some studies mention the accomplishment of pyrolysis in an autoclave using 500 °C for 30 min. Other works mention the completion of pyrolysis in fixed bed reactors using temperatures of 700 or 800 °C. The products obtained in this pyrolysis were oils, gases and solid waste. The solid waste from the pyrolysis contained fiberglass and metal fractions, which could easily be separated. The metallic fraction consisted mainly of copper, calcium, iron, nickel, zinc, and aluminum and low concentrations of precious metals such as gallium, bismuth, silver, and gold [120–122].

Zhou et al. [119, 123] tested a process consisting of vacuum pyrolysis and vacuum centrifugal separation in the treatment of printed circuit board waste. In this study, the PCBs were pyrolysed under vacuum at 600 °C for 30 min in the first step. In the second step, the residue obtained was heated under vacuum until melting of the solder, separating it from the pyrolysis residues by centrifugal force. This way, the oil, gases and solid waste (metals and fiberglass) from the pyrolysis and the solder could be separated. In this case, the solder can be reclaimed and reused directly or serve as good source of lead and tin for other processes.

The use of molten salts (specifically eutectic KOH–NaOH) to dissolve glass and oxides and to destroy plastics, without oxidizing the more valuable metals, has shown to be effective for the recovery of a fraction rich in copper. Analyses of gaseous emissions showed that most of the gas evolved in the process remained trapped in the molten salt. A large amount of hydrogen was produced and could be used as fuel gas or synthesis gas, leading to a favorable energy balance for this new process [124].

In the Noranda process used by some metal recyclers, the materials entering the reactor are immersed in a molten metal bath (1,250 °C), which is churned by a mixture of supercharged air (up to 39 % oxygen). The agitation of the oxidation zone converts impurities including iron, lead and zinc into oxides, which become fixed in a silica-based slag. The copper matte containing precious metals is removed and transferred to the converters. After an upgrading in the converters, liquid blister copper is refined in anode furnaces and cast into anodes with purity of 99.1 %. The remaining 0.9 % contains the precious metals, including gold, silver, platinum and palladium, along with other recoverable metals, such as selenium, tellurium, and nickel. Subsequently, electrorefining of the anodes recovers these marketable metals. The energy cost is reduced by the combustion of plastics and other flammable materials [106].

The processing performed by the company Umicore in Hoboken (Belgium) is focused mainly on the recovery of precious metals. The first step of the recovery of metals is realized in the unit of Precious Metals Operations (PMO). In this unit the smelting is carried out in an IsaSmelt furnace, where the plastics or other organic substances contained in the feed partially substitute the coke as reducing agent and energy source. The smelter separates precious metals in a copper bullion from allmost other metals concentrated in a lead slag, which is further treated at the unit of Base Metals Operations (BMO). The copper bullion is then treated by copper-leaching and electrowinning and the precious metals recovered. The pure metals are recovered in a special metals refinery. In the Umicore plant, a complex flow sheet with several steps, including pyrometallurgical techniques, a hydrometallurgical process, and electrochemical technologies are employed in the recovery of base metals, precious metals, metals from the platinum group and special metals. The furnace has an emissions control system, where gases are cleaned and cooled, with some techniques such as bag house filters, electrofilters, and scrubbers [9].

The main drawback of this type of process is the high investment required by this units, particularly with regard to the need of installing emission control systems.

Cathode ray tubes (CRT) are progressively being replaced by alternative screen technologies, such as liquid crystal display (LCD) screens. The transition from CRT to LCD monitors impacts recycling and disposal systems worldwide. This chapter presents the characteristics and the recycling possibilities of these monitors.

9.3 Monitors

Cathode ray tubes (CRT) are progressively being replaced by alternative screen technologies, such as liquid crystal display (LCD) screens. The transition from CRT to LCD monitors impacts recycling and disposal systems worldwide. This chapter presents the characteristics and the recycling possibilities of these monitors.

9.3.1 Cathode Ray Tubes

Cathode ray tubes monitors are electrostatic deflection oscilloscopes. The development of the cathode ray tube technology began in 1855 with the study of the physical phenomena, and its origin is attributed to numerous inventors and scientists [125]. During the 30s, the popularization of television occurred in several countries [126]. With the emergence of personal computers, this device became a mandatory item to allow interaction between the user and the data processing unit (CPU).

The first CRT monitors used in personal computers were monochrome monitors with low resolution, approximately 140 by 192 pixels. However, the development of new technologies enabled many improvements in the CRT monitor. In 1987, with the appearance of Personal System/2 computer from IBM and the Video Graphics Array (VGA) technology, monitors achieved the resolution of 640 × 480 pixels in color displays of 256 colors [127].

CRT monitors work by the incidence of electrons on a fluorescent panel. A combination of several bright points with three different colors creates the image we see on the monitor screen. The image formation process starts in the monitor's rear part, the electron gun, and ends on the front panel. The whole process can be divided into three parts—the process of formation of the electron beam, the focalization process and the energy conversion process.

The process of formation of the electron beam occurs in the electron gun. It consists of a cathode and a focuser. The electron formation process begins at the cathode of the electron gun. A huge potential difference between the two grids (commonly called acceleration grid and control grid positioned in front of the cathode) and the cathode causes the electrons in the cathode to be pulled away from it. The typical voltage difference of the two grids is 100–1,000 V [128].

To facilitate the electron release process, the cathode is heated between 700 and 1,200 °C. The cathode is usually made of nickel with a thin layer of barium and a strontium oxide surface.

The next step in the image formation process inside the tube is the focusing of the electron beam in the electron gun. The focus and the energy of the electron beam are important to determine the area that the electrons will reach on the phosphor screen and the light intensity of the colors, affecting the resolution of the CRT. The focus of the electron beam is achieved using electrostatic grids or electromagnetic coils. Due to economic features, electrostatic grids are the most used in CRT monitors [129].

After the electron beam exits the cathode, it is focused by a series of three anodes—the pre-acceleration anode, the focusing anode and the acceleration anode. After the electron beam is focused into a central beam, the process of deflection of the electron beam occurs. This process is responsible for guiding the electron beam to a spot on the fluorescent screen. Just like the focusing system, the deflection system can be electrostatic or electromagnetic. In this case, the most used technology in CRT computer monitors and televisions is electromagnetic deflection [130].

The electromagnetic deflection system is composed of a coil placed outside the tube. The coil creates a magnetic field that attracts or repels the electron beam in two directions—horizontally (x axis) and vertically (y axis)—putting them in the right direction to pass through the shadow mask and hit the fluorescent screen in a particular area. Figure 9.2 shows an electromagnetic deflection coil.

Finally, the process of energy conversion occurs. This consists in converting the energy of the electron beam into visible light for the human eye. The phosphorescent materials present on the front glass emit light through fluorescence and phosphorescence processes. Fluorescence is the phenomenon of light emission while the material is being bombarded by electrons, and phosphorescence is the emission of light after the bombardment of electrons [131].

CRT monitors have several components with very different compositions of materials, such as metals, ceramics and plastics. The main components found in CRT monitors are shown in Table 9.2.

The cathode ray tube is a tube with a conical shape formed by three types of glass and the electron gun. It is the main component of the monitor because it is inside the CRT that the image formation process occurs. The cathode ray tube is also the heaviest component of the monitor, reaching 60 % of its total weight.

Fig. 9.2 Electromagnetic deflection system

Table 9.2 CRT main components [132]

CRT component	Materials	Mass (%)
Cathode ray tube unit		
Funnel glass	Glass	14.81
Panel glass	Glass	28.71
Electron gun	Steel, glass, copper, plastic, nickel, strontium	0.82
Shadow mask		3.89
Printed circuit board—main board, electron gun board	Ceramics, epoxy resin, copper, steel, aluminum,	14.34
Polymeric case—front case, base case, main case	ABS Plastic	17.38
Metal parts	Steel, aluminum, brass	6.46
Wires	Copper, plastic	5.65
Rubber parts	Rubber	0.41
Plastic parts	Plastic	2.49
Deflection yoke	Copper, plastic, steel	5.04
Phosphorus coat	Metals	<1

The printed circuit board is responsible for controlling the image formation process of, serving as support for the electrical components. The printed circuit boards are the most complex components of the monitors, due to the great heterogeneity of materials present in their composition.

The polymeric parts, such as the cases and rubbers, represent a small mass percentage and their recycling process is very well established. The common polymers found in CRTs are acrylonitrile-butadiene-styrene co-polymers (ABS), polycarbonates (PC), styrene-butadiene co-polymers (SBR) and high impact polystyrenes (HIPS) [133, 134].

The metals found in CRTs are steel, stainless steel, aluminum, brass and copper. Metals make up a small amount of about 12.11 % of the total. Rare earth and precious metals are present in trace amounts and their recycling routes are not yet well established [135].

During the formation and acceleration of electrons, radiation is emitted from the tube, especially from the front display. The radiation emitted by CRT is in the X-ray wavelength and is caused when the electron beam strikes the phosphor layer [136]. To prevent the escape of radiation from the tube, lead is added to the composition of the glass. Lead has the ability to absorb the ionizing radiation.

The addition of lead occurs in the funnel and neck glass. The panel glass has the addition of barium. Like lead, barium is capable of absorbing a certain amount of radiation because of its density.

The addition of lead to the glass tube is required for radiation levels not to reach values higher than 10 mR/h at a distance of 30 cm from the external surface of the tube. These values are limits imposed by the Food and Drug Administration—FDA in the United States [137] and usually followed by manufacturers of CRT monitors worldwide.

The main environmental risk is associated with the possibility of the lead present in the CRT tube leaching. Several authors point out that the lead present in the glass can leach in various environments and in concentrations higher than the environmental standards [138–140]. There is also a leaching risk of other toxic metals present in the glass, such as barium and strontium, and of cadmium, which can be found in the phosphor layer.

The recycling of CRT waste is a critical step in the recycling of electronic waste. The CRT monitor is the largest and most heavy component of a computer, it's composition is basically glass and the fact that it is a hazardous residue makes the CRT monitor of low interest to the WEEE recycling companies.

The recycling of CRTs is based on closed-loop and open-loop recycling systems. The closed-loop system refers to the recycling of CRTs in the manufacturing chain itself. Here, the old tubes are used in the manufacturing of new CRT tubes. The open-loop system involves the recycling of CRT glass in the manufacture of new products.

Both systems present difficulties/barriers. In the closed-loop system, the recycling of CRTs presents major logistical difficulties. The presence of manufacturers in the country is an important factor that makes this type of recycling logistically and economically feasible. However, many countries have no CRT manufacturers. Even if manufacturers of CRT are present in several countries, the recycling demands would be greater than the demand for new CRT tubes, mainly due to declining sales for new CRTs.

One barrier of open-loop recycling systems is the non-acceptance of CRT glass by glass manufacturers. The variation in the composition of the glass from monitor to monitor hinders the insertion of this type of glass in the manufacture of new glassy objects.

In WEEE recycling companies, the CRT recycling circuit starts with the collection, temporary storage and cleaning of monitors. The problem with collection and storage is that CRT monitors are considered hazardous residue, so its transportation and storage most comply with all the regulations of hazardous waste management.

The dismantling of CRT components is the next step. The CRT monitor is manually opened and its components are removed one by one. Usually, the components are sorted in four types of residue—the cathode ray tube, the polymers, the metals and the printed circuit boards [134].

The CRT is kept intact, without the metallic band and any polymeric adhesive on its surface. The other CRT components follow the conventional recycling systems. Polymers are crushed and sent to recycling companies, metals are sent to smelter companies and the printed circuit boards follow the viable recycling route. Figure 9.3 shows the process flow diagram for the recycling of CRT monitors.

The next step comprises the depressurization and the separation of the glass types. This separation is required because the three types of glass have different chemical compositions and, consequently, different methods of recycling.

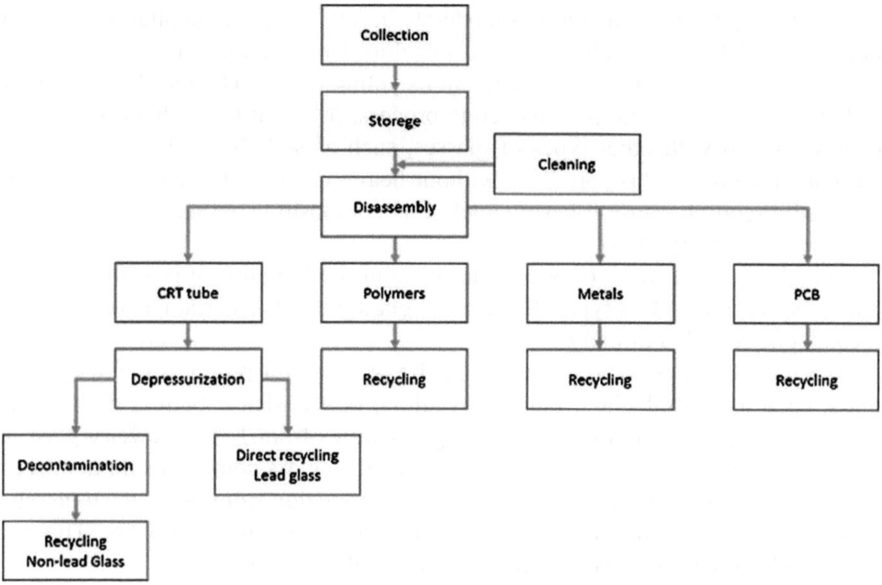

Fig. 9.3 Process flow diagram for CRT recycling

The depressurization is generally carried out by breaking the top of the neck glass or by drilling the anode in the glass funnel [141].

The separation of the tube must be done mechanically because there is the risk of breaking the glass. The two most widely used methods for this purpose are the hot wire method and the diamond saw method. Both techniques are automated with little tube handling. Other technologies involve the laser cutting system, the wet jet separation method and the thermal-shock system method [142].

The hot wire system applies an electric current to a nickel-chromium wire that is coiled around the CRT at the interface between the funnel and the panel glasses. The heat wire melts the glass and an air flow cools down the wire. The thermal shock obtained from the heating and rapid cooling breaks the interface glass. The complete separation between the funnel and the faceplate glass in then completed manually by the operator [143].

The advantage of this method is its low cost and requirement for investment capital. The associated problems are the wrong position of the wire, resulting in an irregular cut, the sharp edges on the interfaces and the possible vapors resulting from the heating [143].

The diamond saw method involves the use of two or more diamond saw blades cutting through the frit interface. This method uses wet or dry conditions. In the wet condition, water is pumped to reduce friction and heat. In the dry condition an airflow is used. This method can have a rotating CRT or rotating diamond saw blades. The correct alignment of the blades and the frit interface is fundamental to obtain a clear cut. The low operating cost and large-scale of operation are the main advantages of this method. A moderate initial investment capital is necessary, however, and the correct alignment must be done by an automated machine [143].

The glass separation enables different recycling routes. The panel glass usually follows the conventional glass recycling process. It is sent to smelters (glass companies), mixed with other types of glasses, such as soda-lime glass, and used in the manufacture of glass products without heavy metals. The panel glass can be recycled directly because it has no lead in its composition, so no decontamination process is necessary [144].

The glass separation process should be monitored to prevent possible contamination between the glass types. If the panel glass presents a certain amount of lead, it cannot be recycled directly.

The funnel and electron gun glass can be recycled for products that use leaded glass, such as crystal glass, radioactive glass, optical dense barium crown glass or jewelry. The addiction of lead in the glass varies from 1 % in jewelry glass, to 65 % in radio shielding glass. In general, however, glass companies do not accept CRT glass because of the variation in glass composition from one CRT to another and the presence of many elements that may interfere in the process. This fact proves to be a barrier in the recycling of leaded glass [145].

It is also possible to recycle leaded glass for products that may contain certain amounts of lead in their composition, such as foam glass, bricks, clays, and piles. Many studies propose the recycling of CRT glass, especially leaded glass, in the ceramics industry [144, 146–148].

The use of leaded glass in the manufacture of glass products with potential direct human contact is restricted because of the possibility of contamination. Whatever the product derived from the leaded glass, therefore, the "encapsulation" of the lead must be ensured.

Ceramic companies can also perform the recycling of leaded glass. Several studies indicate that it is possible to use glass with lead in ceramic products, mainly ceramic construction products, such as windows, doors and insulation materials. However, there are restrictions regarding the possibility of lead leaching from these products [149–152].

The Nulife Glass Company and the Sweepkauusakoski Company employ the process of lead decontamination [153, 154]. They use a furnace and chemicals to remove the lead from the glass. The process is based on the reduction of lead and its removal from the glass matrix. The result is a lead bar and "deleaded" glass. Although few companies in the world use this technology, the process promises a new way to deal with this residue.

9.3.2 Liquid Crystal Displays/Light Emitting Diodes

Liquid crystal displays, known as LCDs, are used for many different purposes, such as cellphones, monitors, televisions, calculators, digital cameras and displays of various equipment.

One of the objectives of this technology is the replacement of the cathode ray tube (CRT) display. The composition of the LCD consists of two glass plates with transparent conductive electrodes, and a liquid crystal layer with a thickness ranging from 4 to 10 μm sandwiched between the conductive electrodes. The composition of the screen also includes polarizers, a diffuser sheet, a reflective sheet, a light guide plate (or perspex), a light source and printed circuit boards [155].

The Austrian Friedrich Reinitzer discovered liquid crystals in 1888. He tested cholesteryl benzoate, an organic fluid, and found that upon heating, the fluid changed of phase from a milky fluid to a more transparent fluid. This was later explained as a transition from an optically and electrically anisotropic fluid to an isotropic fluid [155].

Liquid crystals are an intermediate phase between crystalline solids and isotropic liquids. They combine certain characteristics and properties of the crystal structure with a deformable fluid. Monitors use both its fluidity and the anisotropy associated with its crystalline character.

LCD screens have a typical amount of liquid crystals of 0.6 mg/cm^2 per glass substrate. In a 15″ screen, the total amount of liquid crystal is approximately 400 mg. Liquid crystal mixtures are used, usually containing between 10 and 25 components [156].

The first functional LCD was presented in 1968 by G. Heilmeier RCA, based on the dynamic scattering mode. In 1969, J. Fergason discovered the twisted nemactic field effect in liquid crystals and his company produced the first LCD based on this effect [157].

The idea of driving liquid crystal displays with thin film transistors (TFT) on cross-sections of the array was designed by Bernard Lechner in 1971. Based on this concept, the active matrix monitor was manufactured in 1989. This technology has made high-resolution color monitors possible and the first notebook using this technology was sold in 1990. In 1996, In-Plane Switching (IPS) was first marketed. This technology makes it possible LCD monitors with better contrast and substantially improves the viewing angle [157].

Passive matrix LCD monitors became common in the 80s. They were used in laptops, but the refresh rate of this technology was very slow, which distorted the image. The main advantages were the weight and cost.

In a passive matrix, a grid is used to provide a load to a specific pixel on the monitor. This grid is produced with two glass plates, called substrate. This substrate has columns and lines made of a transparent conductive material, typically indium tin oxide (ITO). In small alphanumeric LCDs with limited information, such as calculators and electronic watches, transparent ITO electrodes are patterned into segments. Each segment is treated individually by electronic system. The rows and columns are connected to integrated circuits that control when the load is send to a particular row or column. To trigger a pixel, the integrated circuit sends a charge to the column specified in one of the substrates and an activated ground on the correct line of the other substrate. The row and column intersect at the designated pixel, and provide the voltage to untwist the liquid crystals at that pixel.

In passive LCD matrix, transparent conductive electrodes are patterned stripes perpendicular to each other in two opposite glass layers. This allows an array of points with M rows and N columns (for example, M $\times$ N pixels) to be treated by external connections M + N. In this system, the voltages that cross each pixel are not controlled independently. One of the disadvantages of this technology is that the passive matrix is best viewed from a frontal perspective, since moving to the side to view the screen from an oblique angle means there is color distortion, darkening or other problems. This technology is therefore only practical for displays containing little information [155].

In monitors with a lot of information content, the large number of elements (pixels) prevents the individual control of each pixel. In the case of active matrix technology, the transistors are recorded on the substrate where each pixel is located. In the active matrix, film transistors (known as TFT) on the glass layer enable each pixel to be individually controlled.

If a passive matrix has a resolution of 640 $\times$ 480 pixels, for example, then 1,120 TFT transistors would be required, whereas for an active matrix 307,200 TFT transistors are needed. It is therefore absolutely necessary to attach the electrode in the substrate. For this resolution, the glass substrate used in the active matrix will contain 307,200 electrodes, while the passive matrix will only have 1,120 electrodes [155].

The production of active matrix displays is more complex, which increases the price in comparison with passive matrix displays of the same resolution.

In an active matrix LCD, the pixel is constantly maintained by the combination of a capacitor and a TFT, which functions as a circuit with short-term memory between updates. This circuit adds cost and complexity to LCDs when compared to CRTs, but when the final cost is considered, one has to take into account that the volume and weight of the CRT creates a lot more transportation, handling and disposal issues than LCDs [158].

In a LCD screen, the source of illumination is located in the back. A reflective sheet is placed, a light guide plate (also known as Perspex, diffuser sheets that allow the light to spread out uniformly and an LCD screen are combined with this light source.

The glass plates have small grooves isolated from each other, each having one electrode connected to a transistor. These grooves have the function of aligning the orientation of helix type liquid crystal molecules.

The glass screen has polarizing films that are perpendicular relative to one another. The first polarizing film that is in contact with the diffuser sheets causes the light emitted inside the screen to be polarized. This light will then only pass through another polarizing film with the presence of the liquid crystal.

To form an image in LCDs, a transparent electrical conductor material is needed to conduct electricity to the liquid crystal and to direct the beam. Indium oxide doped with tin oxide (ITO), a mixture of SnO_2 and In_2O_3, has this characteristic.

The ITO layer combines the features of a solid with good electrical conductivity, resistance to ultraviolet radiation, reflection of infrared radiation, and a transparent, slightly yellowish appearance, according to the concentration of tin oxide, having a transparency of approximately 90 %.

When the liquid crystal molecules are aligned in the form of a helix, they change the direction of the polarized light, which goes through the polarizer film. This configuration is known as "white" mode. When an electric field is applied to the glass screen, the liquid crystals are aligned along the light that passes through the first polarizer, which does not change direction and therefore does not not pass through the second polarizer. This configuration is known as "black" mode.

By controlling the voltage applied between the electrodes present in the ITO layer, it is possible to control the degree of twist of the liquid crystals and thus the intensity of the emitted light of that color. In the front glass plate it is possible to verify the presence of pixels formed by three sub-pixels in red, blue and green. In the rear glass plate the transistors, TFT, can be found.

Each sub-pixel has a transistor that controls it. The transistor acts as a switch that allows the screen to be updated line by line. When a voltage is applied in a specific row, each sub-pixel of this line can receive video data from the top of the screen. Only one line can receive information at a time, but the speed with which this happens for every line is so fast that the screen appears to render a fluid image.

The individual sub-pixels on the screen have red, green and blue filters. The representation of color is based on the principle of additive color: the normal distances from the observer's point of view will effectively create the color mixture of the three subpixels composing each pixel. When the three subpixels are transmitting, the pixel color will be perceived as white. The exact color coordinates of the point (color temperature of the monitor) depends on the transmission and the purity of the colors of the subpixel [155].

The difference between an LCD and an LED monitor is the source of illumination.

The backlight on an LCD active matrix generally uses cold cathode fluorescent lamps (CCFLs) placed at the edge of the panel (in notebooks and on most monitors) or behind the panel (on TVs and some monitors). The backlight CCFLs consist of one or more lamps coated on the inside with red, green, and blue phosphor [155].

The LED array is one of the most popular sources of lighting available. A LED array can illuminate a thin sheet diffuser behind the LCD screen and provide uniform background lighting. The LED backlight is available in red, yellow, blue and other colors, and should be dimmed or turned off by varying the current through the LED. The LED backlight has a much longer duration than other electroluminescent devices with a useful life exceeding 50,000 h [159].

The main manufacturers of LCDs have developed, and are increasingly using, LEDs as a light source to replace CCFLs. Because of its greater production efficiency, LEDs produce less heat, consume less energy, and are superior in almost every way when compared to CCFLs [160].

In the case of In-plane switching (IPS) monitors, the difference is in the liquid crystals, which are aligned horizontally rather than the traditional vertical alignment seen in twisted nemactic monitors.

Unlike the twisted nematic cell, in the IPS cell the electrodes are placed only on one of the glass substrates, resulting in a heterogeneous electrical field in a first order approximation parallel to the substrate surface (equipotential lines are perpendicular to the surface of the glass substrate). The molecules are twisted in the plane of the substrate, leading to a significant reduction in the viewing angle dependence of the intensity of transmitted light when compared to twisted nematic monitors. This improvement in viewing angle dependence, which can be achieved without additional optical compensation film, is the main benefit to the IPS technology and makes it very interesting for monitors.

Another particularity of the IPS cell is a result of the single electrode arrangement and orientation of the layer. While in the twisted nematic cell, equivalent electronic circuit of orientation layer and liquid crystal have a series connection, the structure of the IPS provides a coupling parallel of the orientation layer and the liquid crystal [161].

The basic structure of all LCD monitors consists of:

- Front cover
- Polimeric frame
- LCD screen
- PCB
- Back cover

Figure 9.4 shows a dismantled monitor and its various constituents. Items 1–14 show the various components that form the display.

Item 1: Front of the monitor
Item 2: Monitor without polimeric frame
Item 3: Front of the LCD screen
Item 4: Back of the LCD screen

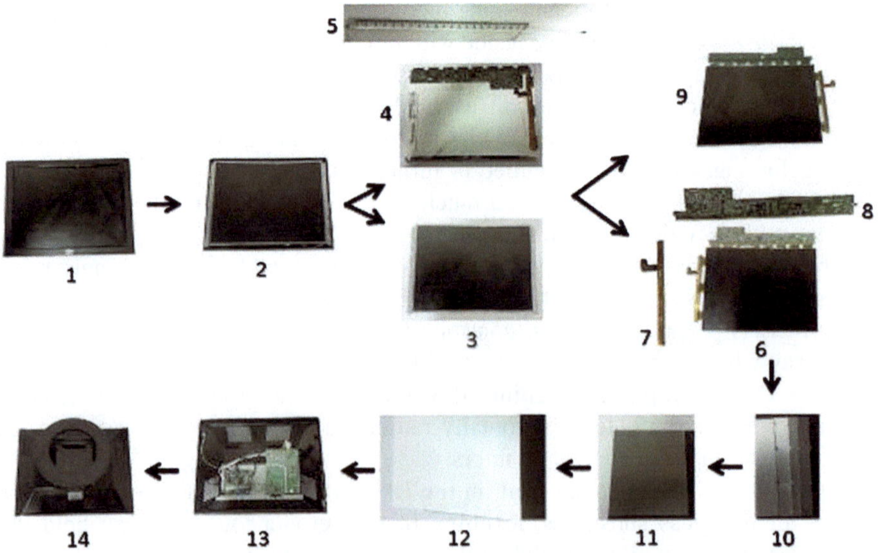

Fig. 9.4 LCD monitor dismantled and its various constituents

Item 5: Fluorescent lamp over a 20 cm ruler
Item 6: Front Screen Glass with PCB
Items 7 and 8: PCB
Item 9: Back of the glass screen
Item 10: Diffuser sheets
Item 11: Perspex
Item 12: Reflective sheet
Item 13: Back cover of the monitor and PCB
Item 14: Back of the monitor

The average composition by weight of an LCD screen is [156]:

- Polimers: 51 %
- Metals: 16 %
- LCD Glass: 22 %
- PCB: 9 %
- Backlight: 0.34 %
- Wire, rubber, laminated plastics and other: 1.7 %

Hunt et al. [162] estimated that more than ten thousand tons of LCDs would be available for recycling in 2010 in the UK alone. This is equivalent to a potential recovery of 9 tonnes of liquid crystal with a market value between 27 and 187 millions of dollars, depending on the quality of the liquid crystal; 900 kg of indium and 8,000 tonnes of optical grade glass could also be recovered. Also according to Hunt et al. incineration is wasteful, environmentally damaging and expensive. It is estimated that 2.5 billion LCDs are approaching their end of life. In combination with conservative forecasts for growth in the sales of LCDs of 16–28 % every five years, this fact indicates that the recovery and recycling of liquid crystal LCDs are of the utmost importance [162].

In recycled flat screen devices, usable components or components containing hazardous substances must be removed manually (such as CCFLs, for example), although this is currently not a common practice since monitors are not designed for easy disassembly. Casings made of plastics and larger metal components must be removed manually. The remaining volume is usually crushed and various separation techniques are used to obtain fractions of materials such as polymers, ferrous and non-ferrous metals.

The latches of an LCD monitor are small and can be inaccessible, including clips, screws and adhesives. Furthermore, the design, including the types and locations where screws and other components are placed, differs between products made by different manufacturers. These barriers make it difficult to design a method of disassembly pattern that can accommodate the different models on the market.

These difficulties mean that many companies prefer to export entire monitors to countries like China, Vietnam, Indonesia, and Malaysia or even to South America and Africa, exposing these devices to a type of recycling that is often technically and environmentally inappropriate.

Some components may be removed from an LCD unit to be separately managed. These include the LCD screen, mercury lamps and PCBs [157].

Glass LCD screens usually do not arouse great interest and may end up as waste in the process. Depending on the recycling company, this waste can be landfilled or incorporated by secondary processors (e.g., recycling of polymers or of smelters companies) [157].

Merck has developed two technology options focused on the recovery of the glass substrate used in LCDs [163]. The glass is typically more than 80 % by weight of the LCD. Currently two types of glass substrate are used for the manufacture of LCDs:

- Soda-lime glass and
- Borosilicate glass

So when the LCDs on the market reach their end of life, the recycling of glass from these displays has to deal with a mixture of two different types of glass. A merger with the selective separation of impurities could be possible [164]. The recovered glass used in this process may then be used in the cement industry or in the ceramics industry [157, 163, 164].

According to the EPA, about 4 mg of elemental mercury is contained in the fluorescent lamps of an LCD unit, which should therefore be handled properly. A mercury lamp can be removed intact by hand. The recovery of the mercury lamp, however, requires special equipment that breaks the lamp and securely separates the glass and mercury and other components. For this, a closed negative pressure process, which prevents the release of mercury, it is necessary. Because of this, only specialized companies are typically able to recover mercury from lamps [157]. It is estimated that between 290 and 480 kg of mercury will ultimately have to be disposed of for all 80 million LCDs in use worldwide in 2010 [165].

The hydrometallurgical methods obtain the mercury and then it can be recovered as a solid product. The humid chemical techniques to treat mercury must meet the following criteria: It must produce a marketable product that can be reintroduced into the production process and the effluents from the applied process should not contain dangerous substances for the environment, or they should be treated untill they reach acceptable levels when they no longer present any danger for the environment [166].

The larger metal components can be removed manually, while the minor, internal components may be recovered through fragmentation and separation. Recyclers recover steel, aluminum, copper, and printed circuit boards.

Steel and aluminum are used in LCD monitors as structural components. Copper is used in the circuit, precious metals and lead (the solder) may also be found on printed circuit boards (PCBs). Larger metal components can be removed manually, whereas the ssmaller or internal components can be recovered by grinding and separation/concentration processes [157].

According to Havlik et al. [93], PCBs are usually crushed and then burned in a furnace, generating a solid residue with concentrated metals. Plastics are burned out to create either volatile compounds or the slag phase with refractory oxides.

During this process, some metals can behave like collectors of other metals. As such, only alloys with unexpected composition are obtained, not pure metals [93].

The precious metals that are found on printed circuit boards of electronic equipment can be recovered by metallurgical industries. The three best known precious metals smelters are the Boliden (Sweden), Noranda (Canada) and Umicore (Belgium) smelters. Since most printed circuit boards are, typically, 70 % non-metallic, 16 % copper, 4 % solder and 2 % nickel, along with iron, silver, gold, palladium and tantalum, approximately 90 % of the intrinsic value of most scrap PCB is in the gold and palladium content. Consequently, traditional reprocessing of circuit boards has concentrated on the recovery of these metals [167].

The polymeric components, such as the epoxy resin of the printed circuit boards, can be used as a source of fuel, in the process of recovery by smelters. Precious metals are also recovered in some countries using rudimentary refining techniques, with less security for workers and environmental controls.

Technically, the recovery of liquid crystals from obsolete or damaged LCDs is possible. However, there are factors that greatly limit the feasibility of recycling, such as:

- the liquid crystals in LCDs are a mixture containing 25 components or more;
- the amount of liquid crystals in each LCD is very small: 0.6 mg per square cm of display size in a layer of about 5 micrometers thick [168, 169];
- the extensive use of solvents is required to recover the liquid crystals, but the use of these solvents is restricted by economic considerations and regulations.

Research by Merck indicates that recovery processes are more expensive than producing the same amount of new material. The recycling of liquid crystal mixtures, therefore, does not appear to be commercially viable at this time [157].

Indium is classified as a "critical metal" by numerous reports because it is a scarce metal that plays an important role in the solar energy industry, in flat screens, and in other new technology applications. For many of these applications, there are no substitutes that offer the same performance [47].

In May 2005, Sharp announced that it has developed a technique to recycle indium from the indium tin oxide (ITO) from the transparent conductive film used in LCD panels. The LCD panel is crushed into shards of glass, which is then is processed in an acid dissolution. The company employs a separation technique for extracting the indium from the indium solution. According to Sharp the technique is quite simple, since it uses commercially available chemicals and does not require a lot of energy, and is still able to recover highly pure indium. Despite being a rare metal, indium should be increasingly used in transparent electrode monitors [157].

Since 2007 more Indium has been produced from primary and secondary raw materials. More than half the world's consumption of indium is for indium tin oxide (ITO) [170]. In 2011, 425 ton of indium were estimated to be used for the production of ITO [171]. The secondary raw material has been the most important waste from the process of sputter deposition of ITO on LCDs, in which only 30 % of ITO is deposited on the glass surface [172]. The sputtering indium waste can be recovered by dissolving in HCl and then precipitating it with NaOH [173].

A cell phone screen contains as little as 5.5 mg of indium per phone [174]. PC monitors contain between 79 and 82 mg of indium, whereas bigger liquid crystal display television screens use 260 mg [175].

Recycling possibilities for indium are limited. Indeed the EOL recycling rate is small, the fraction of secondary metal in the total metal input to metal production, reaches values between 25 and 50 % [176]. Appropriate recycling activities are concentrated in countries like China, Japan and the Republic of Korea, where ITO production and sputtering takes place [47].

Some recycling methods have been patented for the secondary recovery of indium from the ITO residue (used in LCDs). The methods by Lee et al. and Chen et al. [177] are based on leaching and precipitation. As in all precipitation techniques, these methods are limited by the equilibrium solubility of the hydroxide.

Shimofusa et al. [177] patented a method in which indium is recovered by solvent extraction from a nitric acid solution, but this includes impurities such as heavy metals, and the use of nitric acid, which is costly. Clearly, there is still a need to develop a more efficient method to recover indium from waste LCDs [177].

In LEDs, indium can be recovered to an extent in normal pyrometallurgical smelting operations, e.g. during e-waste and copper scrap recycling. This is clear as the carbon line in Ellingham diagram crosses the In_2O_3 at about 720 °C, implying that with the appropriate oxygen partial pressure in the pyrometallurgical reactor, this element can be distributed to phases from which it can be recovered [47].

References

1. Wang R, Xu Z (2014) Recycling of non-metallic fractions from waste electrical and electronic equipment (WEEE): a review. Waste Manag 34(8):1455–1469
2. Widmer R, Oswald-Krapf H, Sinha-Khetriwal D, Schnellmann M, Heinz BH (2005) Global perspectives on e-waste. Environ Impact Assess Rev 25:436–458
3. Oguchi M, Murakami S, Sakanakura S, Kida A, Kameya T (2011) A preliminary categorization of end-of-life electrical and electronic equipment as secondary metal resources. Waste Manag 31:2150–2160
4. UNEP & UNU (United Nations Environment Programme & United Nations University) (2009) Sustainable innovation and technology transfer industrial sector studies—recycling from e-waste to resources. Berlin, Germany
5. Jang YC (2010) Waste electrical and electronic equipment (WEEE) management in Korea: generation, collection, and recycling systems. J Mater Cycles Waste Manag 12:283–294
6. Tanskanen P (2013) Management and recycling of electronic waste. Acta Mater 61:1001–1011
7. Li J, Shrivastava P, Gao Z, Zhang HC (2004) Printed Circuit board recycling: a state-of-the-art survey. IEEE Trans Electron Packag Manuf 27(1):33–42
8. Jang YC, Kim M (2010) Management of used & end-of-life mobile phones in Korea: a review. Resour Conserv Recycl 55:11–19
9. Hagelüken C (2005) Recycling of electronic scrap at Umicore's integrated metals smelter and refinery. In: Proceedings of EMC
10. Cui J, Zhang L (2008) Metallurgical recovery of metals from electronic waste: a review. J Hazard Mater 158:228–256
11. Besiou M, Georgiadis P, Van Wassenhove LN (2012) Official recycling and scavengers: symbiotic or conflicting? Eur J Oper Res 218:563–576

12. Kang HY, Schoenung JM (2005) Electronic waste recycling: a review of U.S. infrastructure and technology options. Resour Conserv Recycl 45:368–400
13. Fisher M (2006) Feedstock recycling technologies in the sustainable recycling of plastics from end-of-life electrical and electronic products. In: IEEE international symposium on electronics and the environment, American Chemistry Council, Arlington, USA
14. Jung SH, Kim SJ, Kim JS (2012) Thermal degradation of acrylonitrile–butadiene–styrene (ABS) containing flame retardants using a fluidized bed reactor: the effects of Ca-based additives on halogen removal. Fuel Process Technol 96:265–270
15. Astm (2000) D5033-00: standard guide for development of ASTM standards relating to recycling and use of recycled plastics
16. Gabriel AP (2012) Caracterização e Reciclagem da Carcaça Polimérica de Monitores de Tubos de Raios Catódicos Pós-Consumo. Universidade Federal do Rio Grande do SulUFRGS, Dissertação de mestrado
17. Tsydenova O, Bengtsson M (2011) Chemical hazards associated with treatment of waste electrical and electronic equipment. Waste Manag 31:45–58
18. Pérez JG, Navarro PF, Castelló A, Cima MFL, Ramis R, Boldo E, Abente GL (2013) Cancer mortality in towns in the vicinity of incinerators and installations for the recovery or disposal of hazardous waste. Environ Int 51:31–44
19. Vilaplana F, Karlsson S, Ribes-Greus A (2007) Changes in the microstructure and morphology of high-impact polystyrene subjected to multiple processing and thermo-oxidative degradation. Eur Polymer J 43:4371–4381
20. Sanches EMS, Felisberti MI, Costa CAR, Galembeck F (2003) Avaliação da Degradação Térmica e Fotooxidativa do ABS para Fins de Reciclagem. Polímeros: Ciência e Tecnologia 13(3):166–172
21. Santos ASF, Agnelli JAM, Manrich S (1999) Estudo da Influência de Resíduos Catalíticos na Degradação de Plásticos Reciclados (Blenda HDPE/PP e PET) Provenientes de Lixo Urbano. Polímeros: Ciência e Tecnologia 9(4):189–194
22. Santana RC, Manrich S (2002) Studies on thermo-mechanical properties of post-consumer high impact polystyrene in five reprocessing steps. Prog Rubber Plast Recycl Technol 18(2):99–110
23. Soriano F, Morales G, Díaz De León R (2006) Recycling of high impact polystyrene in coextruded sheet: influence of the number of processing cycles on the microstructure and macroscopic properties. Polym Eng Sci 46(12):1698–1705
24. Vilaplana F, Ribes-Greus A, Karlsson S (2006) Degradation of recycled high-impact polystyrene. simulation by reprocessing and thermo-oxidation. Polym Degrad Stab 91:21632170
25. Gabriel AP, Santana RMC, Veit HM (2013) Recycling of frames from CRT monitors with different years of manufacture. In: ISWA World Congress, 2013, Vienna
26. Kasper AC, Berselli GBT, Freitas BD, Bernardes AM, Veit HM (2010) Estudo da reciclagem das Blendas PC + ABS e PC + ABS + EPÓXI provenientes de sucatas eletrônicas. Revista Brasileira de Ciências Ambientais 16:26–35
27. Kasper AC, Bernardes AM, Veit HM (2010) Characterization and recovery of polymers from mobile phone scrap. Waste Manage Res 29(7):714–726
28. Taurino R, Pozzi P, Zanasi T (2010) Facile characterization of polymer fractions from waste electrical and electronic equipment (WEEE) for mechanical recycling. Waste Manag 30:2601–2607
29. Juchneski NCF, Scherer J, Grochau IH, Veit HM (2013) Disassembly and characterization of liquid crystal screens. Waste Manag Res 31(6):549–558
30. Ravi V (2012) Evaluating overall quality of recycling of e-waste from end-of-life computers. J Clean Prod 20:145–151
31. Buchanan R (2004) Ceramic materials for electronics, 3rd edn. Marcel Dekker Inc., New YorK
32. Callister WD (2007) Materials science and engineering: an introduction, 7th edn. Wiley, New York
33. Morris PR (1990) A history of the world semiconductor industry. Peter Peregrinus Ltd, London

34. Assefa S (2012) A 90 nm CMOS integrated nano-photonics technology for 25 Gbps WDM optical communications applications. IEEE Int Electr Devices Meeting, Maryland
35. Coombs C (2008) Printed circuits handbook, 6th edn. The McGraw-Hill Ltd, New York
36. Jawitz M (1997) Printed circuit board materials handbook. McGraw-Hill Ltd, New York
37. Wu J (2008) Electrostatic separation for recovering metals and nonmetals from waste printed circuit board: problems and improvements. Environ Sci Technol 42:5272–5276
38. Yokoyama S, Iji M (1997) Recycling of printed wiring boards with mounted electronic parts. In: Proceedings of the 1997 IEEE international symposium, pp 109–114
39. Guo B (2008) A plate produced by nonmetallic materials of pulverized waste printed circuit boards. Environ Sci Technol 42:5267–5271
40. Sohaili J, Muniyandi SK, Mohamad SS (2012) A review on printed circuit boards waste recycling technologies and reuse of recovered nonmetallic materials. Int J Sci Eng Res 3(2):138–144
41. Crompton T (2000) Battery reference book, 3rd edn. Reed Educational and Professional Publishing, Woburn
42. Verhaeghe F (2011) Valorisation of battery recycling slags. 2nd International Slag Valorization Symposium. Leuven, Belgium
43. Niu X, Li Y (2007) Treatment of waste printed wire boards in electronic waste for safe disposal. J Hazard Mater 145:410–416
44. Guo J (2008) Phenolic molding compound filled with nonmetals of waste PCBs. Environ Sci Technol 42:624–628
45. Guo J (2009) Manufacturing process of reproduction plate by nonmetallic materials reclaimed from pulverized printed circuit boards. J Hazard Mater 163:1019–1025
46. Zheng Y (2009) The reuse of nonmetals recycled from waste printed circuit boards as reinforcing fillers in the polypropylene composites. J Hazard Mater 163:600–606
47. Worrell E, Reuter MA (2014) Handbook of recycling: state of the art for practitioners, analysts, and scientists. Elsevier Inc., Amsterdam
48. Hancock P (2011) Fluxing issues—container glass in sanitaryware applications. Asian Ceramics 50–53
49. Tucci A, Esposito L, Palmonari C, Rambaldi E (2004) Use of soda-lime scrap glass as a fluxing agente in porcelain stoneware tile mix. J Eur Ceram Soc 24:83–92
50. Benjarungroj P, Harrison P, Vaughan S, Ren XJ, Morgan M (2012) Investigation of thermally treated recycled glass as a vibratory mass finishing media. Key Eng Mater 496:104–109
51. Cui J, Forssberg E (2003) Mechanical recycling of waste electric and electronic equipment: a review. J Hazard Mater 99:243–263
52. Kasper AC, Berselli GBT, Freitas BD, Tenório JAS, Bernardes AM, Veit HM (2011) Printed wiring boards for mobile phones: characterization and recycling of copper. Waste Manag 31:2536–2545
53. Ladou J (2006) Printed circuit board industry. Int J Hygiene Environ Health 209:211–219
54. Hall WJ, Williams PT (2007) Separation and recovery of materials from scrap printed circuit boards. Resour Conserv Recycl 51:691–709
55. Zeng X, Zheng L, Xie HH, Lu B, Xia K, Chao K, Li W, Yang J, Lin S, Li J (2012) Current status and future perspective of waste printed circuit boards recycling. Procedia Environ Sci 16:590–597
56. Schlummer M, Gruber L, Mäurer A, Wolz G, Van Eldik R (2007) Characterization of polymer fractions from waste electrical and electronic equipment (WEEE) and implications for waste management. Chemosphere 67:1866–1876
57. Lim SR, Schoenung JM (2010) Toxicity potentials from waste cellular phones, and a waste management policy integrating consumer, corporate, and government responsibilities. Waste Manag 30:1653–1660
58. Sepúlveda A, Schluep M, Renaud FG, Streicher M, Kuehr R, Hagelüken C, Gerecke AC (2010) A review of the environmental fate and effects of hazardous substances released from electrical and electronic equipments during recycling: examples from China and India. Environ Impact Assess Rev 30:28–41

59. Tsydenova O, Bengtsson M (2011) Chemical hazards associated with treatment of waste electrical and electronic equipment. Waste Manag 31:45–58
60. Frazzoli C, Orisakwe OE, Dragone R, Mantovani A (2010) Diagnostic health risk assessment of electronic waste on the general population in developing countries' scenarios. Environ Impact Assess Rev 30:388–399
61. Ongondo FO, Williams ID, Cherrett TJ (2011) How are WEEE doing? A global review of the management of electrical and electronic wastes. Waste Manag 31:714–730
62. Wienold J, Recknagel S, Scharf H, Hoppe M, Michaelis M (2011) Elemental analysis of printed circuit boards considering the ROHS regulations. Waste Manag 31:530–535
63. Oliveira CM, Bernardes AM, Gerbase AE (2012) Collection and recycling of electronic scrap: a worldwide overview and comparison with the Brazilian situation. Waste Manag 32:1592–1610
64. Scharnhorst W, Hilty LM, Jolliet O (2006) Life cycle assessment of second generation (2G) and third generation (3G) mobile phone networks. Environ Int 32:656–675
65. Yu J, Williams E, Ju M (2010) Analysis of material and energy consumption of mobile phones in China. Energy Policy 38:4135–4141
66. Long L, Sun S, Zhong S, Dai W, Liu J, Song W (2010) Using vacuum pyrolysis and mechanical processing for recycling waste printed circuit boards. J Hazard Mater 177:626–632
67. Das A, Vidyadhar A, Mehrotra SP (2009) A novel flowsheet for the recovery of metal values from waste printed circuit boards. Resour Conserv Recycl 53:464–469
68. Roth T, Wolf M, Pöhlein M, Van Eldik R (2013) Analysis of flame retardants and elements of concern in printed wiring boards with respect to origin and year of construction. Anal Bioanal Chem 405:7215–7229
69. Norgate T, Haque N (2012) Using life cycle assessment to evaluate some environmental impacts of gold production. J Clean Prod 29–30:53–63
70. Hagelüken C (2006) Improving metal returns and eco-efficiency in electronics recycling. In: Proceedings of the 2006 IEEE international symposium on electronics and the environment pp 218–223
71. Polák M, Drápalová L (2012) Estimation of end of life mobile phones generation: the case study of the Czech Republic. Waste Manag 32:1583–1591
72. Zheng Y, Shen Z, Cai C, Ma S, Xing Y (2009) The reuse of nonmetals recycled from waste printed circuit boards as reinforcing fillers in the polypropylene composites. J Hazard Mater 163:600–606
73. Veit HM, Pereira CC, Bernardes AM (2002) Using mechanical processing in recycling printed wiring boards. JOM 54:45–47
74. Hou S, He X, Yang D, Xu Z (2013) An impact crushing dynamic model of waste printed circuit board particles. Res Chem Intermed 39:3611–3630
75. Zhang S, Forssberg E (1997) Mechanical Separation-Oriented Characterization of Electronic Scrap. Resour Conserv Recycl 21:247–269
76. Guo C, Wang H, Liang W, Fu J, Yi X (2011) Liberation characteristic and physical separation of printed circuit board (PCB). Waste Manag 31:2161–2166
77. Yoo JM, Jeong J, Yoo K, Lee JC, Kim W (2009) Enrichment of the metallic components from waste printed circuit boards by a mechanical separation process using a stamp Mill. Waste Manag 29:1132–1137
78. Duan C, Wen X, Shi C, Zhao Y, Wen B, He Y (2009) Recovery of metals from waste printed circuit boards by a mechanical method using a water médium. J Hazard Mater 166:478–482
79. Veit HM, Pereira CC, Hörlle LF, Bernardes AM (2002) Recycling of metals from PCB by dense medium separation processes. In: Recycling and waste treatment in mineral and metal processing: technical and economic aspects, Lulea, Sweden
80. Svoboda J, Fujita T (2003) Recent developments in magnetic methods of material separation. Miner Eng 16:785–792
81. Li J, Xu Z, Zhou Y (2007) Application of corona discharge and electrostatic force to separate metals and nonmetals from crushed particles of waste printed circuit boards. J Electrostat 65:233–238

82. Li J, Lu H, Xu Z, Zhou Y (2008) Critical rotational speed model of the rotating roll electrode in corona electrostatic separation for recycling waste printed circuit boards. J Hazard Mater 154:331–336
83. Lu H, Li J, Guo J, Xu Z (2008) Movement behavior in electrostatic separation: recycling of metal materials from waste printed circuit board. J Mater Process Technol 197:101–108
84. Zhang S, Forssberg E, Arvidson B, Moss W (1998) Separation mechanisms and criteria of a rotating eddy-current separator operation. Resour Conserv Recycl 25:215–232
85. Zhang S, Forssberg E, Arvidson B, Moss W (1998) Aluminum recovery from electronic scrap by High-Force® eddy-current separators. Resour Conserv Recycl 23:225–241
86. Veit HM, Diehl TR, Salami AP, Rodrigues JS, Bernardes AM, Tenório JAS (2005) Utilization of magnetic and electrostatic separation in the recycling of printed circuit board scrap. Waste Manag 25(1):67–74
87. Veit HM, Bernardes AM, Ferreira JZ, Tenório JAS, Malfatti CF (2006) Recovery of copper from printed circuit boards scraps by mechanical processing and electrometallurgy. J Hazard Mater B137:1704–1709
88. Tuncuk A, Stazi V, Akcil A, Yazici EY, Deveci H (2002) Aqueous metal recovery techniques from e-scrap: hydrometallurgy in recycling. Miner Eng 25:28–37
89. Kinoshita T, Akita S, Kobayashi N, Nii S, Kawaizumi F, Takahashi K (2003) Metal recovery from non-mounted printed wiring boards via hydrometallurgical processing. Hydrometallurgy 69:73–79
90. Kim EY, Kim MS, Lee JC, Jha MK, Yoo K, Jeong J (2008) Effect of cuprous ions on Cu leaching in the recycling of waste PCBs, using electro-generated chlorine in hydrochloric acid solution. Miner Eng 21:121–128
91. Jha MK, Choubey PK, Jha AK, Kumari A, Lee JC, Kumar V, Jeong J (2012) Leaching studies for tin recovery from waste e-scrap. Waste Manag 32:1919–1925
92. Yazici EY, Deveci H (2013) Extraction of metals from waste printed circuit boards (WPCBs) in $H2SO_4$–$CuSO_4$–NaCl solutions. Hydrometallurgy 139:30–38
93. Havlik T, Orac D, Petranikova M, Miskufova A, Kukurugya F, Takacova Z (2010) Leaching of copper and tin from used printed circuit boards after thermal treatment. J Hazard Mater 183:866–873
94. Park YJ, Fray DJ (2009) Recovery of high purity precious metals from printed circuit boards. J Hazard Mater 164:1152–1158
95. Zhang Y, Liu S, Xie H, Zeng X, Li J (2012) Current status on leaching precious metals from waste printed circuit boards. Procedia Environ Sci 16:560–568
96. Jing-Ying L, Xiu-Li X, Wen-Quan L (2012) Thiourea leaching gold and silver from the printed circuit boards of waste mobile phones. Waste Manag 32:1209–1212
97. Gurung M, Adhikari BB, Kawakita H, Ohto K, Inoue K, Alam S (2013) Recovery of gold and silver from spent mobile phones by means of acidothiourea leaching followed by adsorption using biosorbent prepared from persimmon tannin. Hydrometallurgy 133:84–93
98. Birloaga I, De Michelis I, Ferella F, Buzatu M, Veglio F (2013) Study on the influence of various factors in the hydrometallurgical processing of waste printed circuit boards for copper and gold recovery. Waste Manag 33:935–941
99. Oh CJ, Lee SO, Yang HS, Ha TJ, Kim MJ (2003) Selective leaching of valuable metals from waste printed circuit boards. Air & Waste Manag Assoc 53:897–902
100. Ha VH, Lee J, Jeong J, Hai HT, Jha MK (2010) Thiosulfate leaching of gold from waste mobile phones. J Hazard Mater 178:1115–1119
101. Syed S (2012) Recovery of gold from secondary sources-A review. Hydrometallurgy 115–116:30–51
102. Scott K, Chen X, Atkinson JW, Todd M, Armstrong RD (1997) Electrochemical recycling of tin, lead and copper from stripping solution in the manufacture of circuit boards. Resour Conserv Recycl 20:43–55
103. Fleming CA (1992) Hydrometallurgy of precious metals recovery. Hydrometallurgy 30:127–162

104. Sulka GD, Jaskula M (2002) Study of the kinetics of the cementation of silver ions onto copper in a rotating cylinder systemfromacidic sulphate solutions. Hydrometallurgy 64:13–33
105. Choo WL, Jeffrey MI (2004) An electrochemical study of copper cementation of gold(I) thiosulfate. Hydrometallurgy 71:351–362
106. Cui J, Zhang L (2008) Metallurgical recovery of metals from electronic waste: a review. J Hazard Mater 158:228–256
107. Cortina JL, Meinhardt E, Roijals O, Marti V (1998) Modification and preparation of polymeric adsorbents for precious-metal extraction in hydrometallurgical processes. React Funct Polym 36:149–165
108. Nakbanpote W, Thiravetyan P, Kalambaheti C (2000) Preconcentration of gold by rice husk ash. Miner Eng 13:391–400
109. Chand R, Watari T, Inoue K, Kawakita H, Luitel HN, Parajuli D, Torikai T, Yada M (2009) Selective adsorption of precious metals from hydrochloric acid solutions using porous carbon prepared from barley straw and rice husk. Miner Eng 22:1277–1282
110. Zhang H, Ritchie IM, La Brooy SR (2004) The adsorption of gold thiourea complex onto activated carbon. Hydrometallurgy 72(3–4):291–301
111. Wang J, Bai J, Xu J, Liang B (2009) Bioleaching of metals from printed wire boards by Acidithiobacillus ferrooxidans and Acidithiobacillus thiooxidans and their mixture. J Hazard Mater 172:1100–1105
112. Xiang Y, Wu P, Zhu N, Zhang T, Liu W, Wu J, Li P (2010) Bioleaching of copper from waste printed circuit boards by bacterial consortium enriched from acid mine drainage. J Hazard Mater 184:812–818
113. Ilyas S, Anwar MA, Niazi SB, Ghauri MA (2007) Bioleaching of metals from electronic scrap by moderately thermophilic acidophilic bactéria. Hydrometallurgy 88:180–188
114. Santos FHS (2010) Resíduos de origem eletrônicaSérie Tecnologia Ambiental 57. Rio de JaneiroCETEM/MCT
115. Zhu N, Xiang Y, Zhang T, Wu P, Dang Z, Li P, Wu J (2011) Bioleaching of metal concentrates of waste printed circuit boards by mixed culture of acidophilic bactéria. J Hazard Mater 192:614–619
116. Hoffmann JE (1992) Recovering precious metals from electronic scrap. JOM 44(7):43–48
117. Lee JC, Song HT, Yoo JM (2007) Present status of the recycling of waste electrical and electronic equipment in Korea. Resour Conserv Recycl 50(4):380–397
118. Sum EYL (1991) The recovery of metals from electronic scrap. JOM 43(4):53–61
119. Zhou Y, Qiu K (2010) A new technology for recycling materials from waste printed circuit boards. J Hazard Mater 175:823–828
120. De Marco I, Caballero BM, Chomón MJ, Laresgoiti MF, Torres A, Fernández G, Arnaiz S (2008) Pyrolysis of electrical and electronic wastes. J Anal Appl Pyrol 82:179–183
121. Hall WJ, Williams PT (2007) Separation and recovery of materials from scrap printed circuit boards. Resour Conserv Recycl 51:691–709
122. Quan C, Li A, Gao N, Dan Z (2010) Characterization of products recycling from PCB waste pyrolysis. J Anal Appl Pyrol 89:102–106
123. Zhou Y, Wu W, Qiu K (2010) Recovery of materials from waste printed circuit boards by vacuum pyrolysis and vacuum centrifugal separation. Waste Manag 30:2299–2304
124. Flandinet L, Tedjar F, Ghetta V, Fouletier J (2012) Metals recovering from waste printed circuit boards (WPCBs) using molten salts. J Hazard Mater 213–214:485–490
125. Miller FP, Vandome AF, Mcbrewster J (2009) Cathode ray tube. Alphascript Publishing, ISBN 6130087101
126. Medical discoveries. Cathode ray tube (CRT). Available at http://www.discoveriesinmedicine.com/Bar-Cod/Cathode-Ray-Tube-CRT.html. Accessed 13 Jun 2014
127. Polsson K (2014) Chronology of personal computers. Available at http://pctimeline.info. Accessed 13 Jun 2014

128. Rahim Z (1989) Texas instruments understanding the operation of a CRT Monitor. Available at http://www.ti.com/lit/an/snla017/snla017.pdf. Accessed 13 Jun 2014
129. Keller P (1991) The cathode-ray tube, technology, history, and applications, Edn 1. Palisades Press
130. Holms A (2005) An introduction to the cathode ray tube. CRT report. University of California. Available at http://www.writing.ucsb.edu/faculty/holms/2E_CRT_report_2.pdf. Accessed 13 Jun 2014
131. Ozawa L (1990) Cathodoluminescence: theory and applications, Edn 1. Kodansha
132. Lee CH (2000) Management of scrap computer recycling in Taiwan. J Hazard Mater A73:209–220
133. Socolof ML (2005) Environmental life-cycle impacts of CRT and LCD desktop computer displays. J Clean Prod 13:1281–1294
134. Nnorom IC, Osibanjo O, Ogwuegbu MOC (2011) Global disposal strategies for waste cathode ray tubes. Resour Conserv Recycl 55:275–290
135. Dexpert-Ghys J (2009) Re-processing CRT phosphors for mercury-free applications. J Lumin 129:1968–1972
136. Parr G, Davie H (1959) The cathode ray tube and its applications, Edn 3. Butler
137. Food and Drug Adminstration U.S. (2012) Code of Federal Regulations Title 21Part 1020 performance standards for ionizing radiation emitting products. Available at http://www.accessdata.fda.gov/scripts/cdrh/cfdocs/cfcfr/CFRSearch.cfm?fr=1020.20. Accessed 13 Jun 2014
138. Lee CH, Hsi CS (2002) Recycling of scrap cathode ray tubes. Environ Sci Technol 36:69–75
139. Townsend T (2004) RCRA toxicity characterization of computer CPUs and other discarded electronic devices. State University System of Florida, Florida Center for Solid and Hazardous Waste Management, p 79
140. Musson SE (2000) Characterization of lead leachability from cathode ray tubes using the toxicity characteristic leaching procedure. Environ Sci Technol 34:4376–4381
141. Menad N (1999) Cathode ray tube recycling. Resour Conserv Recycl 26:143–154
142. Qingbo Xu, Li Guangming, He Wenzhi, Huang Juwen, Shi Xiang (2012) Cathode ray tube (CRT) recycling: current capabilities in China and research progress. Waste Manag 32:1566–1574
143. Herat S. Recycling of cathode ray tubes (CRTs) in electronic waste. Available on http://www98.griffith.edu.au/dspace/bitstream/handle/10072/22550/52113_1.pdf;jsessionid=15070CD1298A959A9B6928D111C2057A?sequence=1
144. Icer (2004) Materials recovery from waste cathode ray tubes (CRTs). The Waste & Resources Action Programme. ISBN: 1-84405-077-7
145. Lee C, His C (2002) Recycling of scrap cathode ray tubes. Environ Sci Technol 36:69–75
146. Bernardo E (2009) Glass based stoneware as a promising route for the recycling of waste glasses. Adv Appl Ceram 108(1):2–8
147. Méar F, Yot P, Ribes M (2006) Effects of temperature, reaction time and reducing agent content on the synthesis of macroporous foam glasses from waste funnel glasses. Mater Lett 60:929–934
148. Guo HW, Gong YX, Gao SY (2010) Preparation of high strength foam glass–ceramics from waste cathode ray tube. Mater Lett 64:997–999
149. Project code: GLA15-006 (2004) Materials recovery from waste cathode ray tubes (CRTs), The Waste and Resource Action Programme. ICER, UK
150. Hreglich S, Falcone R, Vallotto M (2001) The recycling of EOL panel glass from TV sets in glass fibres and ceramic productions. In: Proceedings of the international symposium on recycling and reuse of glass cullet, pp 123–134
151. Weiand W, Merz R, Bonk K (1992) Glaze additives: 30 years ago to today. Tile Brick 1(8):14–17
152. Piccagliani V (2002) Preliminary study of the reuse of CRT glass in the ceramic field, materials Eng. Degree thesis, Modena and Reggio Emilia University
153. Nulife Glass (2013) Recycling CRTs from televisions & computer screens. Available at http://www.nulifeglass.com/. Accessed 13 Jun 2014

154. Sweep Kuusakoski (2013). Available at http://www.sweeepkuusakoski.co.uk/. Accessed 12 Jun 2014
155. den Boer W (2005) Active matrix liquid crystal displays. Elsevier, Amsterdam
156. Mester A, Fraunholcz N, Van Schaik A, Reuter MA (2005) Characterization of the hazardous components in end-of-life notebook display. In: Light metals 2005, Canada
157. King CountySolid Waste Division (2008) Flat panel displays: end of life management report. Available on http://your.kingcounty.gov/solidwaste/takeitback/electronics/documents/FPDReport.pdf. Accessed 10 Jun 2014
158. Reinhard E, Ward G, Pattanaik S, Debevec P (2006) High dynamic range imaging: acquisition, display and image-based lighting. Elsevier, Amsterdam
159. Smith JR (2005) Programming the PIC microcontroller with MBASIC. Elsevier, Amsterdam
160. Chen RH (2011) Liquid crystal displays. Wiley, Hoboken
161. Heckmeier M, Lüssem G, Tarumi K, Becker W (2002) Liquid crystals for active matrix displays. Available on http://www.emdmillipore.com/showBrochure/200908.037.ProNet.pdf. Accessed 05 Jun 2014
162. Hunt AJ, Clark JH, Breeden SW, Matharu AS, Ellis C, Goodby JW, Bottomley JA, Cowling SJ (2008) Extraction of liquid crystals from flat panel display devices using both liquid and supercritical carbon dioxide. Available on http://www.isasf.net/fileadmin/files/Docs/Barcelona/ISASF%202008/PDF/Key%20notes/KN_PR_5.pdf. Accessed 05 Jun 2014
163. Martin R, Simon-Hettich B, Becker W (2004) New EU Legislation (WEEE) compliant recovery processes for LCDs. Presented at 11th international display workshop, IDW, Germany
164. Becker W, Simon-Hettich B, Honicke P (2003) Toxicological and ecotoxicological investigations of liquid crystals and disposal of LCDs. In: Presented at 11th international display workshop, IDW, Germany
165. Ryan A, O'Donoghue L, Lewis H (2011) Characterizing components of liquid crystal displays to facilitate disassembly. J Clean Prod 19:1066–1071
166. Buzatu M, Milea NB (2008) Recycling the liquid crystal displays. UPB Sci Bull Ser B 70(4):93–102
167. Weee & Hazardous Waste (2004) AEA Technology Environment. http://archive.defra.gov.uk/environment/waste/producer/electrical/documents/weee-hazwaste.pdf. Accessed 13 Jun 2014
168. Becker W, Simon-Hettich B, Honicke P (2007) Toxicological and ecotoxicological investigations of liquid crystals. Available on http://www.emdmillipore.com/showBrochure?id=200810.029. Accessed 13 Jun 2014
169. LCD Industries Research Committee (LIREC) (2013) Japan electronics and information technology industries association environment and safety activities. Available on http://www.jeita.or.jp/english/about/2012/profile_en_2012.pdf. Accessed 13 Jun 2014
170. Chou WL, Huang YH (2009) Electrochemical removal of indium ions from aqueous solution using iron electrodes. J Hazard Mater 172:46–53
171. Talens PL, Villalba MG (2013) Material flow analysis of scarce metals: sources, functions, end-uses, and aspects for future supply. Environ Sci Technol 47(6):2939–2947
172. Tolcin AC (2008) Minerals Yearbook-Indium. Available on http://minerals.usgs.gov/minerals/pubs/commodity/indium/myb1-2008-indiu.pdf. Accessed 13 Jun 2014
173. Ryoshi K, Hattori Y, Kobayashi H (2008) Method of manufacturing powder containing indium oxide as a main component. Japanese Patent JP2008137825
174. Yoshioka K, Nakajima T (1994) Source of total factor productivity for Japanese manufacturing industry: 1962–1988. Keiko University, Tokyo
175. Buchert M, Schueler D (2009) Critical metals for future sustainable technologies and their recycling potential, Sustainable innovation and technology transfer industrial sector studies. Unep, Nairobi
176. UNEP (2011) Recycling rates of metals—a status report. Available at http://www.unep.org/resourcepanel/Portals/24102/PDFs/Metals_Recycling_Rates_110412-1.pdf. Accessed 03 Set 2014
177. Virolainen S, Ibana D, Paatero E (2011) Recovery of indium from indium tin oxide by solvent extraction. Hydrometallurgy 107:56–61

Chapter 10
Batteries

Felipe Costa Hashimoto Bertin, Denise Crocce Romano Espinosa, Jorge Alberto Soares Tenório, Eduardo Hiromitsu Tanabe, Daniel Assumpção Bertuol, Edson Luiz Foletto and Ana Javorsky da Costa

The high variety of applications for batteries, from flashlights to vehicles, demands may different battery chemistries and materials so as to meet the wide range of power, energy, size, weight, safety and cost requirements [1]. Among these different types of batteries, primary batteries (nonrechargeable), such as the alkaline and zinc varieties, and rechargeable batteries, such as lead/acid secondary batteries can be found. Portable secondary batteries are the most widely used in electrical and electronic equipments. The recycling processes of these portable rechargeable batteries (NiCd, NiMH and LIBs), along with the recycling of primary zinc-manganese dioxide systems, are going to be discussed in this chapter.

10.1 Nickel–Cadmium (NiCd) Batteries

Felipe Costa Hashimoto Bertin, Denise Crocce Romano Espinosa, Jorge Alberto Soares Tenório

Polytechnic School, USP, São Paulo, Brazil
e-mail: espinosa@usp.br

Waldemar Jungner developed the Nickel-Cadmium alkaline battery in 1899 and because their components were expensive at the time, its use was limited to applications where overall costs were not a key factor. By 1932, Schlecht and Ackerman had invented the porous nickel electrode, which increased the surface area of active laid up materials, inducing a better reaction kinetics. This fact meant a significant improvement, making sealed batteries as efficient as they are today [2, 3]. These sealed batteries were made possible when Newman managed, in 1947, to recombine the gaseous form generated during charge conditions (overcharged condition necessary in 10 %, producing oxygen), replacing the ventilated designs and allowing this battery to expand its market share [3, 4].

H.M. Veit and A. Moura Bernardes (eds.), *Electronic Waste*, Topics in Mining, Metallurgy and Materials Engineering, DOI 10.1007/978-3-319-15714-6_10

Even though some slight modifications have been made to the design of these batteries over time, according to producer's estimates the average composition of Ni-Cd batteries was Fe (40 %), Ni (22 %), Cd (15 %), Plastic (5 %), KOH (2 %), others (16 %) in 2010 [5], although according to some research papers, the actual composition consists in double those average percentages for Ni and Cd plus other metals such as Cobalt with an average composition inferior to 3 % [6].

A part from knowing the average metallic concentration, it is also necessary to consider which are the main metallic compounds present in batteries, because changes in the recycling process may be needed to efficiently extract each metallic element from its compound. As such, the material to be recycled can be described as [7]:

- Cathode made of metallic Cd
- Anode made of NiOOH (which is where the Co is found) and
- KOH electrolytes impregnating an absorbent polymer sheet

The discharge reaction can be described by Eq. 10.1 [8]:

$$2NiOOH + Cd + 2H_2O = 2Ni(OH)_2 + Cd(OH)_2 \quad (10.1)$$

The present configuration of a Ni-Cd Battery is similar to those of alkaline batteries, which allowed them to be a viable replacement for primary batteries, but they have some limitations that need to be considered, such as the relatively low energy density (compared to new systems) and the accumulated memory effect that inefficient charge-discharge cycles can produce [2]. Another drawback is the fact that new policies and requirements all over the world, such as the National Solid Waste Policy (PNRS) in Brazil, now limits the use of metals with high toxicity such as Cd [9, 10].

Cadmium is considered highly toxic because it is carcinogenic and it can cause death for fetuses in early stages of development, among other health and environmental problems [11]. Most of the products that contain Cd have therefore appeared on the crosshairs of many regulations and policies seeking to limit its concentration.

Despite all its advantages compared to other types of batteries, however, this system has a faster charging rate and high resistance to overcharge and over discharge cycles. It also can be manufactured in several different shapes and sizes and is highly economic in terms of cost per cycle, which meant this battery had a 50 % market share worldwide in 2001 [4, 12].

These facts prevent the Ni-Cd batteries from being totally replaced by other types of batteries, so they are expected to be found in urban household wastes. To prevent them from contaminating groundwater reservoirs near landfills, therefore, recycling is necessary.

Because of this legal and environmental necessity, several recycling processes emerged over the last 20 years using basically three methods based on two types of operations: a physical separation, by manually or by unity operations of mining treatment, and a chemical extraction/recovery of the metals through pyrometallurgical and/or hydrometallurgical routes [13, 14].

10.1.1 Manual Sorting

This separation is performed by manually separating different (or similar) types of batteries by size, composition and manufacturer to minimize contamination during further processes. Because this sorting can present a dangerous working environment through the presence of toxic vapors and other dangerous substances, manufacturers usually stimulate consumers to correctly discard each battery at a specific location when possible [5].

10.1.2 Component Separation by Unity Operations of Mineral Treatment

Even when there is pre-sorting of batteries, it is necessary to use operations commonly used in mining facilities to separate and concentrate materials of interest before further recovery processes are carried out ant the unwanted materials is properly discarded. Considering that manual sorting usually cannot separate elements so as to recycle them, these unity operations are the first step prior to any chemical stage, with the mechanical process being of relatively low cost compared to the rest of the operation [13].

Processes commonly used are [15]:

- Comminution to reduce grain size and increase surface area for posterior reactions
- Grain size separation for re-processing by comminution or for further processing
- Magnetic separation, which efficiently recovers iron scraps

10.1.3 Pyrometallurgical Route

Because it uses high temperatures, pyrometallurgical processes are usually associated with high energy consumption and the production of gases that need to be treated before release into atmosphere [16]. One advantage is that due to the high temperature and the low boiling point of mercury, this toxic metal can be recovered at the beginning of pyrometallurgical processes from the solid waste of batteries, decontaminating it for further processing [17].

Espinosa et al. summarize two processes for recycling Ni-Cd batteries: by evaporating Cd in an open furnace recovering the Cd as cadmium oxide, and distilling it from the Ni-Cd batteries into pure Cd powder and into a high-content nickel alloy on a closed furnace with controlled atmosphere. In the distillation process, the monoxides are decomposed into oxides of these metals and there is a further decomposition of cadmium oxide into vapor of metallic cadmium by the partial reduction of Cd vapor pressure or by the presence of an reducing agent [18, 19].

Three examples in operation today are the Snam-Savam process in France and the Sab Nife process in Sweden, which are totally based on pyrometallurgical methods for the recovery of Ni-Cd batteries, and the Inmetco process in United States, which accepts Ni-Cd batteries in its process, although it was originally designed to recycle dust from electric arc furnaces [16, 18, 19].

10.1.4 Hydrometallurgical Route

Hydrometallurgy consists of acidic or alkaline leaching processes. It aims to extract the metals present in the solid waste into soluble forms and then to selectively extract each one by different methods, such as: liquid-liquid extraction (solvent extraction), selectively precipitating them and/or by electrowinning [13, 18].

Among the hydrometallurgical routes, Velgosová et al. [20] studied the bioleaching of Ni-Cd battery waste in acid solutions, the process consists of using a bacterium that oxides Fe^{2+} to Fe^{3+}. The Fe^{3+} will then induce metals into the metallic form to be soluble in solution for further recovery.

The main example of current processes using hydrometallurgy is the one from the Toegepast-Natuurwetenschappelijk Onderzoek (TNO), which was specially designed for the recycling of Ni-Cd batteries in order to recover Cd, Ni and Fe. The process also allows the recycling of small alkaline batteries [13, 21].

10.2 Nickel Metal Hydride (NiMH) Batteries

Eduardo Hiromitsu Tanabe, Daniel Assumpção Bertuol

Laboratory of Environmental Process (LAPAM), Center of Technology – UFSM, Santa Maria, Brazil
e-mail: dbertuol@gmail.com

Since the development of lead-acid batteries by Gaston Plante in 1860, and the development of nickel cadmium batteries by Thomas Alva Edison and Waldemar Junger in 1901, no other new system of rechargeable batteries gained commercial importance until the development of the NiMH battery. Despite the efforts to develop and improve rechargeable NiMH batteries, which began in 1950, it was only in 1980 that these batteries gained commercial importance due to the improvement of the alloy initially used ($LaNi_5$) [22, 23].

10.2.1 Characteristics of Nickel Metal Hydride Batteries—NiMH

Technological progress has brought with it the need for lighter, compact and durable rechargeable batteries to be used in electronics and mobile phones. That is, batteries with an increase in their autonomy. Nickel-metal hydride batteries (NiMH) were developed with this in mind. These batteries have almost twice the energy density of nickel-cadmium batteries (NiCd) and a similar operating voltage, with the benefit that their constituents are less toxic to the environment [22].

With a basic structure identical to the nickel-cadmium batteries, the nickel-metal hydride batteries are composed of layers containing nickel hydroxide, which act as the positive electrode, and an alloy with hydrogen absorbing properties as the negative electrode [24].

The hydrogen absorbing alloys has an absorption capacity equivalent to a thousand times its own volume, generating metal hydrides. These alloys combine a metal **A**, which releases heat in the formation of hydrides, with a metal **B**, whose hydrides absorb heat. Depending on how metals A and B are combined, their alloys are classified into the following types: **AB** (TiFe, etc.), $\mathbf{AB_2}$ ($TiNi_2$, $ZnMn_2$, etc.), $\mathbf{AB_5}$ ($LaNi_5$, etc.) e $\mathbf{A_2B}$ (Mg_2Ni, etc.) [25]. For economic reasons, the La present in $\mathbf{AB_5}$ is often replaced by a rare earth alloy known as *Mischmetal* [26]. *Mischmetal* is a mixture of rare-earth elements, usually composed mainly of Ce, associated with La, Nd, Pr and others, in the proportions in which they occur naturally in minerals [27]. Regarding the charge and discharge efficiency and durability of the different alloys cited, only $\mathbf{AB_5}$ and $\mathbf{AB_2}$ have practical applications.

The task of storing hydrogen raises great challenges. If we consider its use as automotive fuel, for example, then a tank capable of storing 4 kg—a sufficient amount of pressurized hydrogen for an autonomy of 400 km—would be too big to fit on a car. Liquefied hydrogen would occupy less space, but would still be too bulky. Metal hydrides, such as LaNi5, could be capable of releasing hydrogen in the required amount [27].

The nickel-metal hydride batteries are assembled in different formats, such as the cylindrical type (Fig. 10.1a), button type (Fig. 10.1c) or prismatic type (Fig. 10.1d). Cell phones are powered by prismatic or cylindrical batteries connected in series and arranged inside the protective packaging in different spatial arrangements according to the manufacturer. Figure 10.1b shows the assembly of a battery composed by cylindrical accumulators.

The different layers present in the batteries are wrapped in a spiral or in a plate. The two electrodes are separated by a polymer membrane, usually polypropylene or polyamide, amid an alkaline electrolyte, usually 30 % KOH, which allows the movement of electric charges. The whole assembly is coated with a metal or plastic housing. In cylindrical batteries, the positive and negative plates are coiled spirally, and in prismatic batteries they are assembled in layers. For safety, the system has a safety valve, which starts working in the case of high pressure to avoid an explosion [22, 28].

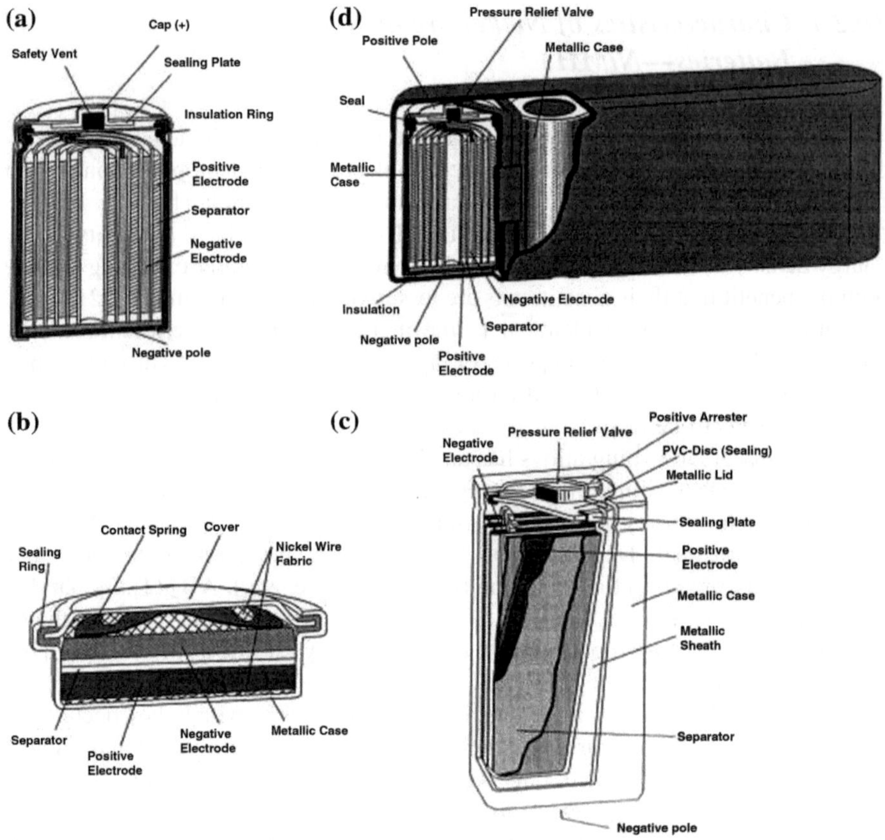

Fig. 10.1 Structures of some NiMH batteries [28]

The operation of cell phone batteries occurs through charging and discharging processes. During charging, the provided energy favors the generation of hydrogen, which migrates from the positive to the negative electrode, charging the battery. These reactions occur after completing the battery charge [28].

The following electrochemical reactions occur in rechargeable batteries: ($\mathbf{M} = \mathbf{AB_5}$ or $\mathbf{AB_2}$) [22, 24, 28]:

$$\text{Overall reaction} \quad Ni(OH)_2 + M \underset{\text{Discharge}}{\overset{\text{Charge}}{\rightleftarrows}} NiOOH + MH \tag{10.2}$$

$$\text{Reaction on the positive electrode} \quad Ni(OH)_2 + OH^- \rightleftarrows NiOOH + H_2O + e^- \tag{10.3}$$

$$\text{Reaction on the negative electrode} \quad M + H_2O + e^- \rightleftarrows MH + OH^- \tag{10.4}$$

10.2.2 Recycling NiMH Batteries

In general, NiMH batteries consist of approximately 36–42 % Ni, 22–25 % Fe, 8–10 % Rare earth metals (REM) and 3–4 % Co. Because of the economic value of metals such as nickel, cobalt, and rare earth metals, efficient recycling processes should be developed to achieve the goals of waste minimization. Recycling this type of waste may bring environmental profits, in addition to economic benefits. Although several processes for recycling of batteries have been developed, until recently, the existing recycling processes for NiMH batteries aimed to recover Ni from a mixture of different batteries, or to use them in stainless steel production as a cheap nickel source. The rare earth elements were lost in the smelter slags [29].

A recycling process usually begins with the mechanical processing step. Mechanical processing operations such as grinding, sieving, magnetic, electrostatic, and gravity separations are employed to liberate the electrodes materials and concentrate valuable metals [30–32]. Using such methods as a pretreatment step before using more expensive pyrometallurgical or hydrometallurgical processes will reduces the amount of impurities in these processes inputs which in turn reduces their cost and increases their selectivity [33].

Müller and Friedrich developed a pyrometallurgical recycling process in which the NiMH batteries are dismantled and melted in an electric arc furnace, producing a nickel–cobalt alloy and a slag phase highly enriched with RE-oxides [29].

Several research groups have developed hydrometallurgical methods for the recovery of nickel, cobalt and rare earth elements from NiMH batteries [34]. Bertuol et al. [35], developed a hydrometallurgical process in which more than 98 wt% of the rare earths were recovered as sulfate salts by dissolution with sulfuric acid, followed by selective precipitation at pH 1.2 using sodium hydroxide. Various studies were conducted using solvent extraction, [30, 36–40], adsorption [41] and electrowinning [42].

Industrially, the recycling of REMs from NiMH battery scrap is in its infancy. In 2011, Umicore (Hoboken, Belgium) and Rhodia (La Rochelle, France) jointly developed a unique pyro-/hydrometallurgical process for the recycling of REMs from rechargeable battery scrap. Not only nickel metal hydride batteries, but also Li-ion batteries can be recycled in this facility. Battery modules are fed directly into an ultra high-temperature smelter without any pretreatment (except for the dismantling of large battery cases). Battery production scrap and slag forming agents are also added to create three output fractions [1, 34]:

- Metal alloy—Co, Ni, Cu, Fe
- Slag fraction—Al, Li, Mn, REM
- Gas emission—flue dust (only fraction to be landfilled)

The oxide slags can be processed to recover the lithium and to produce rare-earth concentrates that are subsequently used as a feed in the rare-earth separation plant of Rhodia in La Rochelle (France). Recently, Honda Motor Co., Ltd. and the Japan

Metals & Chemicals Co., Ltd. announced that they will establish a recycling plant to extract REEs from used nickel-metal hydride batteries collected from Honda hybrid vehicles at Honda dealers inside and outside of Japan. However, the recycling technology and process has not been disclosed [1, 34].

10.3 Lithium Ion Batteries

Edson Luiz Foletto[1], Eduardo Hiromitsu Tanabe[1], Daniel Assumpção Bertuol[1], Ana Javorsky da Costa[2]

[1]*Laboratory of Environmental Process (LAPAM), Center of Technology – UFSM, Santa Maria, Brazil*
e-mail: dbertuol@gmail.com

[2]*Programa de Pós-Graduação em Engenharia de Minas, Metalúrgica e de Materiais – PPGE3M, Universidade Federal do Rio Grande do Sul – UFRGS, Porto Alegre, RS, Brazil*
e-mail: anajavorsky@gmail.com

In the past decade, rechargeable lithium batteries have dominated the market of high power storage systems in portable electronic devices such as mobile phones, video, audio technology, and personal computers due to its many advantages over other batteries (NiCd and NiMH batteries), such as a high energy density (120 Wh/kg), high battery voltage (up to 3–6 V), longevity (500–1,000 cycles) and wide temperature range (20–60 °C) [43–45].

There are two types of lithium-ion batteries, primary (not rechargeable) and secondary (rechargeable). The first primary lithium-ion batteries have been sold by Sony since 1991 [46–49], while secondary, rechargeable, lithium-ion batteries (LIBs) have less than 10 years of commercial development [50, 51]. There is an important difference between primary and secondary lithium-ion batteries (LIBs). Primary lithium-ion batteries use metallic lithium as cathode and contain no toxic metals, but there is the possibility of fire if metallic lithium is exposed to moisture while the cells are corroding. Secondary lithium-ion batteries do not contain metallic lithium [52].

Although widely used in various portable electronic devices, lithium-ion batteries have only recently entered the market for electrical vehicles (EV) and plug-in electric vehicles (PHEV) to reduce CO_2 emissions. In this type of application, the lithium-ion battery cannot only provide high energy densities, but also deliver high power densities in order to be considered as a potential replacement of conventional gasoline engines [53, 54].

10.3.1 *Constituents of Rechargeable Lithium-Ion Batteries (LIBs)*

In general, a secondary lithium ion battery essentially consists of a cathode and an anode electrode, a separator and an organic electrolyte [45, 51]. Approximate percentages of each LIB component are given in Table 10.1 [55].

Cathode materials can be divided into three groups, namely (1) lithium based metal oxides, such as $LiCoO_2$ (2) transition metal phosphates, such as $Li_3V_2(PO_4)_3$ and $LiFeO_4$ and (3) spinels such as $LiMnO_4$ [53, 54, 56]. In the anodes carbon is the typical material used in lithium ion batteries [53]. The cathode electrode is usually linked together with aluminum foil while the anode electrode is linked with copper foil, both adhered to a substrate by a polymeric binder such as polyvinylidene fluoride (PVDF).

The PVDF binder will not react with most acids and bases, most oxidants and halogens, and wil not dissolve in such organic reagents as fatty hydrocarbons, aromatic hydrocarbons, aldehydes and alcohol at room temperature, and will dissolve only partly in some especial ketones and ethers [45, 52].

Finally, the electrodes are immersed in electrolytes that offer high ionic conductivity. The most commonly used electrolyte is based on lithium salts in aprotic solvents, for example $LiPF_6$ in organic solutions, such as ethylene carbonate (EC) and diethyl carbonate (DEC), while the separator is made of micro porous polypropylene (PE) [45, 53].

Figure 10.2 shows a typical configuration of a secondary lithium-ion battery. During the discharge process, lithium ions stored in the anode are oxidized to their ionic form and transported through the electrolyte to the cathode. The electrons flow into the storage particle in the cathode through the external circuit. The lithium ions recombine with the electrons and remain in the storage particle in the cathode. During the charging process the reverse occurs, with lithium leaving storage particles in the cathode and moving into storage particles in the anode. The process of removing or inserting lithium into the storage particles is known as intercalation. The performance of the anode and cathode is generally quantified in terms of capacity per unit mass or per unit of the electrode material, where it is

Table 10.1 Average material components of lithium ion batteries [55]

Components	wt%
Cathode material	25
Anode material	17
Electrolyte	10
Casing	25
Copper electrode foil	8
Aluminium electrode foil	5
Separator	4
Other	6

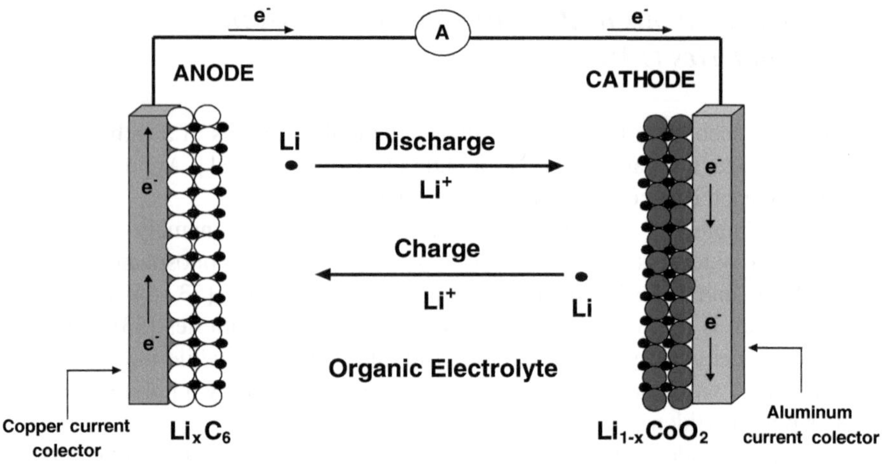

Fig. 10.2 Typical lithium ion battery configuration [59]

the total number of ampere-hours that can be withdrawn from a fully charged cell under specified discharge conditions [53, 57].

The chemical reactions in the two electrodes can be simply expressed by Eqs. 10.5 and 10.6 [52, 58]:

$$\text{The catodic reaction: } 6C + xLi^+ + xe^- \leftrightarrow C_6Li_x \tag{10.5}$$

$$\text{The anodic reaction: } LiCoO_2 \leftrightarrow Li_{(1-x)}CoO_2 + xLi^+ + xe^- \tag{10.6}$$

where the forward direction is the charge reaction and the reverse is the discharge reaction. The energy is stored in these batteries through the movement of lithium ions from the cathode to the anode (charge process of the battery) or vice versa (discharge process) according to Eq. 10.7 [52]:

$$\text{Charge and discharge reaction: } LiCoO_2 + 6C \leftrightarrow Li_{(1-x)}CoO_2 + C_6Li_x \tag{10.7}$$

Lithium-ion battery technology is developing fast. Much effort has been put into fabricating each component. Crucial battery parameters, such as specific energy and power, safety and cycle life, clearly depend on the performance of the electrode and electrolyte.

10.3.1.1 Cathode Materials

As mentioned earlier, in a lithium-ion battery the cathode material can be categorized into three different types: lithium metal oxides, spinels and lithium metal phosphates. The performance of a cathode material is largely governed by the structural stability and conductivity of the cathode. Furthermore, a cathode should be able to accommodate reversible intercalation and deintercalation of lithium just like the anodes. Lithium ion diffusivity therefore plays a significant role in cathodes as well [53].

In **Lithium metal oxides**, the cathode material most commonly used in lithium ion batteries is $LiCoO_2$. $LiCoO_2$ offers significant advantages with respect to ease of synthesis, high working potential and excellent cycle ability at room temperature. Although $LiCoO_2$ is a successful cathode material, alternatives are being developed to lower costs and improve stability. $LiCoO_2$ is not as stable as other potential electrode materials and can experience performance degradation or failure when overcharged. Cycling at voltages above 4.2 V has lead to a rapid degradation in capacity, primarily attributed to structural changes and the dissolution of cobalt in the electrolyte. The other limitation with $LiCoO_2$ is that it is restricted because of a relatively low diffusion coefficient between 10^{-12} and 10^{-9} cm^2s^{-1}. This is particularly problematic when building cathodes for high performance. Finally, $LiCoO_2$ is highly toxic and expensive [53, 60, 61].

In a recent study, coating $LiCoO_2$ nanoparticles with CuO showed significant improvements in performance because of the advantages associated with nanoscale cathode materials and the structural stability induced by coatings. The CuO prevents the dissolution of Co into the electrolyte when charged to a voltage above the conventional 4.2 V [53, 62].

The toxicity and cost of $LiCoO_2$ has been addressed by opting for alternative transition metal oxide cathodes, such as $LiNiO_2$ and $LiMnO_2$. $LiNiO_2$ is lower in cost and has a higher energy density, but it is less stable and less ordered when compared to $LiCoO_2$ [53, 60]. On the other hand, $LiMnO_2$ showed considerable success as a viable metal oxide cathode, with capacities in excess of 250 mA h g^{-1} and a stable cycling ability up to 4.4 V [53].

Vanadium oxides and vanadium derivatives have been considered to be other promising candidates for cathode materials. Vanadium oxides such as orthorhombic V_2O_5 and monoclinic LiV_3O_8 have been used as cathode materials. These electrodes have high capacities, but relatively low voltages (typically 3 V or less) when compared to the compounds discussed above [60].

In **Spinel cathodes**, a promising material for battery manufacture is the lithium manganese spinel ($LiMn_2O_4$). $LiMn_2O_4$ has become the most attractive cathode material for transportation and large-scale batteries due to its low cost, environmental friendliness, good structural stability, and much improved safety. However, during their early stages of development, spinel cathodes encountered limitations with respect to capacity fade associated with such factors as structural changes, manganese dissolution, distortions and microstrains. Dissolution of manganese into the electrolyte has been effectively controlled by oxide coating the spinel electrodes. The addition of Ni in the spinel electrode effectively reduces microstains, thereby offering a better cycle life [53, 59, 60].

The small dimensions of nano-sized active materials can greatly increase the electrode–electrolyte contact area, facilitating charge-transfer and diffusion kinetics during Li insertion/extraction, and offering improved tolerance to structural distortion. These nanomaterials are therefore promising to achieve attractive power capability and cycling stability by using nanostructured $LiMn_2O_4$ and $LiMn_2O_4$/

carbon nanocomposites. Carbon materials [carbon nanotubes (CNT) and graphene nanosheets (GNS)] are added to $LiMn_2O_4$ to make $LiMn_2O_4$/carbon nanocomposites. These composites can provide excellent conducting support for the $LiMn_2O_4$ electrode, suppress the aggregation of nanoparticles, and minimize structural degradation due to volume changes induced by lithium insertion/extraction [63].

Another promising class of cathode material are the **Phosphates** of the olivine family. The phosphate most commonly used for the cathode is $LiFePO_4$. This material has a reasonably high capacity (170 mA h g^{-1}), good stability both at room temperature and at high temperature, excellent cycling performance, high safety, low raw material cost, no pollution, and a rich source of raw materials. However, the low intrinsic electronic conductivity of $LiFePO_4$ and the low diffusion coefficient of Li^+ are the main shortcomings that limit its application in industry. To overcome these drawbacks, three strategies can be adopted: (i) carbon coating to increase the electronic conductivity; (ii) dispersing metal powders or metal oxide coating; (iii) doping metal ions to increase the intrinsic electronic conductivity. Conductivity can be improved by incorporating conductive carbon in the cathode structure, doping with supervalent cations (Mg^{2+}, Al^{3+}, Ti^{4+}, Zn^{4+}, Nb^{5+} or W^{+6}) or using nanostructure $LiFePO_4$ [53, 59, 60, 64]. Other phosphates used for cathodes in lithium ion batteries include $LiMnPO_4$ and $LiCoPO_4$. These materials have open circuit voltages higher than $LiFePO_4$, but have lower capacities [60].

10.3.1.2 Anode Materials

Carbon has typically been favoured as anode material in lithium ion batteries because of its excellent cycling ability and long cycle life. Graphite is commonplace among commercial lithium ion batteries because of its low expansion during lithium insertion. This low expansion is directly linked to a battery's ability to maintain its charge capacity after many charge-discharge cycles. The lithium insertion capacity of graphite (372 mA h g^{-1}) is relatively low, and lithium ion cells stand much to gain if this value is increased. In addition, the diffusivity of lithium ions in graphite is exceptionally high, thereby facilitating rapid charging and discharging [53, 65].

Carbon nanotubes (CNTs) are promising candidates for use as anode material in lithium ion batteries. Carbon nanotubes have demonstrated to be very effective buffering components, and can serve as the backbone in nanostructured anode materials since they can alleviate the degradation of the structural integrity that often results from the significant volume change associated with the charging and discharging process. In addition, the highly conductive carbon nanotubes offer enhanced electronic transport in these nanostructured anode materials. Single-walled carbon nanotubes can have reversible capacities anywhere from 300 to 600 mA h g^{-1}, which means significantly higher capacities can be achieved than with graphite, a widely used material for battery electrodes. More importantly,

CNTs can also be used in composites with metallic or oxide nanoparticles to improve the electrochemical performance of these particles. On the other hand, some limitations of carbon nanotubes are their irreversible lithium ion capacity and lack of a stable voltage as the battery discharges [65].

The **Silicon** anode system (Si) has the highest gravimetric capacity (4,200 mA h g^{-1}), which is much higher than that of graphite (372 mA h g^{-1}). In addition, it has a relatively low discharge voltage (the average delithiation voltage of Si is 0.4 V). However, such electrodes don't hold the structure during charge–discharge cycles because of a large change in volume of the active material during the cycles. Moreover, Si anodes have low electron transfer characteristics due to their inherently low conductivities, and because the cycling ability of amorphous Si is also limited by the operating voltage [53, 66].

The nanostructuring of silicon electrodes has helped overcome the aforementioned limitations to a considerable extent. Silicon/Carbon Nanofiber (Si/CNF) nanocomposite materials have shown to be potentially high performance anodes for rechargeable lithium ion batteries. The carbon nanofiber is hollow, with a graphitic interior and turbostratic exterior. Amorphous silicon layers were uniformly coated via chemical vapor deposition on both the exterior and interior surfaces of the CNF. The resulting Si/CNF composites were tested as anodes for Li ion batteries and exhibited capacities near 800 mAh g^{-1} for 100 cycles [67].

More recently, **Graphene** has attracted attention as a potential anode material in lithium ion batteries. It has advantages over other carbonaceous materials because graphene has relatively low costs and is easy to access. Anodes that use only graphene, however, can only provide a relatively low lithium storage capacity and unstable/poor cycle performance. A number of graphene based composites, using various carbonaceous materials and nanosized metal/oxides as templates, have therefore been developed to enhance the lithium storage capacity and the cycling performance of the graphene anode [68]. Many graphene/metal oxide composites, including Co_3O_4, NiO, Mn_3O_4, MnO, TiO_2, SnO_2 and MoO_2 have been synthesized as anodes for lithium ion batteries [69].

Titanium dioxide (TiO_2) is one of the most attractive anode materials for LIBs due to its high capability, long cycle life, high safety, low cost, nontoxicity, and the fact that it is environmentally benign. **Nano-sized TiO_2** is another attractive anode material, which has fast lithium insertion/extraction ability and lower volume change during the charge/discharge process. Finally, spinel lithium titanium oxides, such as **$Li_4Ti_5O_{12}$** (LTO), have been studied intensively for use as an anode material in lithium secondary batteries. Contrary to the conventional graphite commercialized as an anode active material, LTO could reduce the lithium dendrite problem ("zero strain" material) through its lithium insertion/extraction process, which occurs at a comparatively higher voltage of about 1.55 V (vs. Li/Li^+). LTO is also known to have some advantages associated with lower volume change during charge/discharge to provide superior cycle performance, chemical and electrochemical stability, and excellent thermal characteristics to improve battery safety [70, 71].

However, $Li_4Ti_5O_{12}$ still suffers from its extremely low electronic conductivity and unsatisfactory lithium ionic conductivity, which seriously limits the rate-capability. Lattice doping with aliovalent cations, or simply introducing a second phase with high electronic conductivity, such as Ag, Cu, C, CNTs, and graphene, can both increase the electronic conductivity. The preparation of nanostructured $Li_4Ti_5O_{12}$ has proven to be an effective way to improve the ionic conductivity of electrodes [71].

Tin oxide (SnO_2) has been intensively studied as anode material for lithium ion batteries (LIBs) because of its low cost, low potential for lithium-ion intercalation and higher theoretical capacity (782 mAh g^{-1}). However, the large volume change (up to 300 %) associated with alloying and dealloying causes critical mechanical damage to the electrode, resulting in loss of capacity and rechargeability [72, 73].

To overcome this problem, various types of SnO_2 nanostructures, such as nanotubes, nanowires and nanorods, have been investigated and were found to have enhanced electrochemical performance. They provide more reaction sites, shorter Li^+ ion diffusion lengths and considerable kinetic enhancement. However, the high costs involved in synthesizing nanostructured materials and the relatively low capacity retention still need to be addressed for their future commercial application [73].

10.3.1.3 Electrolytes

An Electrolyte for lithium-ion batteries should be able to dissolve and dissociate into the solvent system, and the Li ions should be able to diffuse in the system with high mobility. In most cases the typical liquid electrolyte solutions used for lithium-ion batteries are **Organic solutions**, such as ethylene carbonate (EC), ethyl methyl carbonate (EMC), dimethyl carbonate (DMC), diethyl carbonate (DEC), propylene carbonate (PC) based solutions containing **lithium salts,** such as lithium hexafluorophosphate ($LiPF_6$), lithium perchlorate ($LiClO_4$), lithium tetrafluoroborate ($LiBF_4$), and lithium hexafluoroarsenate ($LiAsF_6$) [74, 75].

When considering electrolyte salts, $LiClO_4$ remains stable up to 5.1 V and has high conductivity in EC/DMC. Lower impedance has been reported for solid electrolyte interfaces (SEI) formed on anode surfaces. However, $LiClO_4$ is a strong oxidant and reacts with most organic species. $LiBF_4$ remains stable to up to about 5 V versus (Li/Li^+), but its application is limited by the low conductivity in EC/DMC. The preferred salt has been $LiPF_6$ because of its rapid dissolution in carbonate and its low cost. However, due to the flammability of the solvent and the limited chemical and thermal stability of $LiPF_6$, its use poses serious safety risks and strongly reduces the battery's operating temperature range [76].

Solid materials for application as battery electrolyte include a large variety of materials, such as gels, organic polymers, organic–inorganic hybrids, and inorganic materials. Inorganic electrolytes include three categories: crystalline, glass and glass–ceramic electrolytes. Oxide and sulfide compounds have been widely studied as inorganic electrolytes.

Among the **Crystalline electrolytes**, crystalline phosphates with NASICON (Na Super Ionic Conductor) compounds such as $Li_{1+x}Al_xGe_{2-x}(PO_4)_3$ (LAGP) and $Li_{1+x}Ti_{2-x}Al_x(PO_4)_3$ (LTAP), are good Li^+ ions conductors. Perovskite $Li_{0.5-3x}La_{0.5+x}TiO_3$ (LLT) has a high level of bulk conductivity of 10^{-3} S cm^{-1}. LATP and LLT electrolytes have the element Ti in the crystalline structure, and this transition metal is easily reduced by the negative lithium metal electrode; It is therefore difficult for these electrolytes to be used in lithium metal batteries.

Recently, garnet-type $Li_7La_3Zr_2O_{12}$ (LLZ) crystals have been attracting much attention because of their high conductivity of 3×10^{-4} S cm^{-1} and high chemical stability against the negative lithium electrode. All-solid-state rechargeable lithium metal batteries with the LLZ electrolyte showed a good cycleability [77].

The conductivities of the garnet-type oxides are still lower than NASICON-type phosphates and perovskite-type oxides, but they show the following advantages: one is a small grain boundary resistance, and the other is stability against lithium metal. The second advantage of garnet-type oxides is compatibility with the lithium anode [78]. Very recently, it has been found that $Li_{10}GeP_2S_{12}$ with a different crystal structure than the thio-LISICON phase has a higher conductivity of 1.2×10^{-2} S cm^{-1} [77].

In **Glass electrolytes**, most oxide glass electrolytes show a low conductivity at room temperature, but some glasses with high Li^+ ion concentration, such as lithium ortho-oxosalts, have relatively high ion conductivity (10^{-6} S cm^{-1}). LiPON (Lithium Phosphorous Oxynitride) amorphous thin-film also has the same level of conductivity. However, reducing the electrolyte thickness reduces resistance, so LiPON has been used in thin-film batteries [77].

Sulfide glasses with high Li^+ ion concentration show high conductivity. In the system $Li_2S–P_2S_5$, the glasses with P_2S_5 between 20 and 30 % have a conductivity of over 10^{-4} S cm^{-1}, depending on the degree of crystallization [77]. The conductivity of the glass–ceramics (i.e. after crystallization) are higher than those of the corresponding glasses at lower temperatures, which indicates that the crystalline phase has a higher conductivity and lower activation energy than the amorphous phase [78].

In the system $Li_2S–P_2S_5$, $Li_7P_3S_{11}$ or $Li_{3.25}P_{0.95}S_4$ was crystallized from the glasses at the compositions of 70 or 80 mol% of Li_2S, respectively and the prepared glass-ceramics showed high conductivities of over 10^{-3} S cm^{-1} [78].

Solid polymer electrolytes are promising candidates, for they have several advantages over their liquid counterparts. The advantages of these electrolytes include no internal shorting, no leakage of electrolytes, and non-combustible reaction products at the electrode surface, in contrast to the liquid electrolytes [79].

The most commonly used solid polymer electrolyte in lithium-ion conducting electrolytes is poly(ethylene oxide) (PEO). PEO is effective in solvating lithium salts, which are added to provide the lithium-ion conduction. However, at room temperature the ionic conductivity of Li salts dissolved in PEO is limited because the highly symmetrical repeating units in PEO tend to crystallize. Crystalline regions in PEO (m.p. approx. 65 °C) are not available for ion transport, and conductivity is therefore limited to the amorphous regions of PEO (glass transition temperature Tg approx. 55 °C) [78, 80].

Thus, one approach to increase the conductivity is to add a plasticizer, such as succinonitrile (SN), polysquarate (PPS) or ethylene carbonate (EC)/propylene carbonate (PC), which reduces crystallization and thus increases conductivity [78]. Recently, PEO, a conducting salt (lithium bis (trifluoromethansulfonyl)imide; LiTFSI) and a ionic liquid (N-methyl-N-butylpyrrolidinium bis (trifluoromethansulfonyl)imide; $PYR_{14}TFSI$) were mixed and processed at elevated temperatures under dry conditions. Through this process, ternary composites PEO/IL/LiTFSI/ with high ionic conductivity were obtained in the form of thin films [80].

Polymer gel electrolytes (PGEs) prepared by soaking porous polymer membranes in liquid electrolytes (organic solvent or ionic liquids) provide safer and lighter high performance batteries with leak proof construction when compared to traditional liquid electrolytes. In the last decade, many PGPEs have been developed based on monopolymers or copolymers, including poly(vinylidenefluoride) (PVdF) and its co-polymer poly(vinylidene fluoride-hexafluoropropylene) (PVdF-HFP), poly(methyl methacrylate) (PMMA), poly(acrylonitrile) (PAN). PGES based on these materials show high ionic conductivity (in the range of 10^{-4}–10^{-3} S cm^{-1} at room temperature) and good electrochemical stability [81, 82].

By adding organic carbonates like ethylene carbonate, diethyl carbonate, or propylene carbonate to the polymer gel electrolytes, the conductivity of the gels can be dramatically increased so that lithium mobility is more and more favored [83].

Another approach for electrolytes in lithium-ion batteries is the use of room temperature ionic liquids (RTIL), which can be used to replace the traditional organic solvent-based electrolytes. Ionic liquids such as N-n-butyl-N-ethylpyrrolidiniumN,N-bis(trifluoromethane)sulfonamide ($Py_{24}TFSI$) or N-butyl-N-ethylpiperidinium N,N-bis(trifluoromethane sulfonimide) ($PP_{24}TFSI$) are neither volatile nor flammable, which improves the safety of the battery during use. The conductivities of the RTIL-based PVdF-HFP electrolytes are similar to those using $LiFP_6$ and a carbonate solvent [78].

10.3.1.4 Separator

A separator is an inactive, but critical component in a lithium-ion battery. It separates the positive and negative electrodes and provides the ionic conduction through the liquid electrolyte that fills in its porous structure [84]. The key properties/characteristics for separator membranes are thickness, permeability, porosity/pore size, wettability and chemical, thermal and mechanical stability [85]. An ideal separator should have zero ionic resistance, but in practice a low ionic resistance can be achieved by a high porosity. The porosity of a conventional polyolefin separator usually falls between 30 and 55 % [86].

The most commonly used commercial separators in lithium-ion batteries are manufactured from polyolefins, predominantly polyethylene (PE) or polypropylene (PP).

They have many advantages in terms of practical application in commercial lithium-ion batteries. However, due to the poor thermal shrinkage and weak mechanical strength of polyolefin separators, it is difficult to fully ensure electrical isolation between electrodes. Moreover, their intrinsically hydrophobic character and low porosity have often raised serious concerns about insufficient electrolyte wettability, which could directly impair the ion transport through polyolefin separators [84, 87]. Among the various approaches used to achieve this goal, the incorporation of nonwoven fabrics that consist of multi-fibrous layers has drawn considerable attention due to their excellent thermal properties, high porosity, and cost competitiveness. polyacrylonitrile (PAN), polyvinylidene fluoride (PVDF), and polymer blends of polyimide with porosities ranging from 65 to 85 %, have been prepared with a relatively slow electrospinning process. Those spun separators have shown negligible shrinkages at elevated temperatures and enabled good ionic conductivities. However, electrospinning is a relatively high cost process and the polymers used in these work can be plasticized by liquid electrolytes, leading to reduced mechanical properties [86].

10.3.2 Recycling LIBs Batteries

These batteries are a rich source of metals of great commercial interest. Around 36 % of the battery's metal compositions is cobalt. Lithium corresponds to about 6 % of its metal content. Additionally, these batteries contain around 13 % copper and 9.5 % aluminum [88]. The most recently developed processes for recycling LIBs tend to combine different routes and techniques in order to reach a better product and optimize environmental performance.

Thermal processes are usually associated with the production of metallic alloys. Castillo et al. [89] reported that a solid residue of spent LIBs and consisting of Fe, Co and Ni hydroxides and some traces of $Mn(OH)_3$, was introduced into a furnace to eliminate carbon and organics, and the alloy could then directly undergo metallurgical beneficiation.

Saeki et al. [90] developed an effective mechanochemical process consisting of co-grinding $LiCoO_2$ with PVC in a planetary ball mill in air to form Li and Co-chlorides, with subsequent leaching with water in order to extract Co and Li [91].

Zhang et al. [58] studied the leaching of $LiCoO_2$ by the use of H_2SO_3, $NH_4OH{\cdot}HCl$ and HCl. Leaching efficiency of Co is highest in hydrochloric acid and increases with temperature. Zhang et al. [24] also developed a hydrometallurgical technology where anode materials were leached with hot hydrochloric acid where cobalt is extracted from the solution by PC-88A. The raffinate was concentrated by evaporation and lithium carbonate is precipitated by adding $NaCO_3$. Lee and Rhee [92] studied the leaching of $LiCoO_2$ by the use of H_2SO_4 and HNO_3 to substitute HCl with the addition of H_2O_2 as a reducing agent. By the addition of H_2O_2, leaching efficiency increased.

Contestabile et al. [93] developed a laboratory scale hydro-pyrometallurgical process in which the electrode materials were effectively separated by crushing and grinding, and then immersed in N-methyl pyrolidene (NMP) at 100 °C for 1 h to dissolve the binder and liberate active materials from their supporting substrate. The powder was then leached with hydrochloride acid where cobalt was precipitated and recovered as $Co(OH)_2$ by addition of one equivalent volume of a 4 M NaOH solution. The precipitation of cobalt hydroxide begins at pH 6 and is considered completed at pH 8.

In the Kawakami [94] recycling process, lithium dissolves in water and lithium hydroxide is separated from the solution by precipitation and filtration.

The Hsinch process [95] uses membrane electrolysis in LiBs recycling, enabling the recycling of copper and cobalt as cathodes and a precipitate of lithium carbonate.

Mishra et al. [96] used a bio-hydrometallurgical process to recycle spent LiBs. They used *Acidithiobacillus Ferrooxidans* bacteria to leach cathode waste powder in presence of sulfur and iron in ferrous state. Results revealed that a culture of ferrooxidans could metabolize the sulfur and produce sulfuric acid to leach metals indirectly from the LIBs.

Chen et al. [97] suggested the use of vacuum-aided pyrometallurgy to recover metallic cobalt and Li_2O from spent LiBs battery scrap [33].

The Cirimat process is a hydro-pyrometallurgical process for the recycling of LiBs, and it includes digestion of the electrode materials with hot nitric and sulfuric acid, followed by selective precipitation of lithium and manganese as hydroxides. The remaining solid is incinerated to remove carbon and organic materials, producing a metal alloy [98].

Myoung et al. [99] reported that by leaching waste $LiCoO_2$ using a nitric acid solution, cobalt ions can be transformed into $Co(OH)_2$ on a titanium electrode and cobalt oxide is then obtained via a dehydration procedure. Shen [100] investigated the cobalt recovery process from spent LIBs by sulfuric acid leaching and electrowinning. Recovery of cobalt is more than 93 %. Lupi and Pasquali [101] developed a hydrometallurgical process to recycle lithium-ion and lithium-polymer batteries with lithium/cobalt/nickel oxide as cathode material. Nickel was recovered by electrowinning. In order to gain maximum value, Lain [102] developed a process based on mechanical shredding, electrolyte extraction, electrode dissolution and cobalt reduction, which aimed to recover every component from the cell at room temperatures. Aqueous lithium hydroxide was used as electrolyte, and the current collector was graphite.

Solvent extraction has been successfully employed in order to recover metals from a mixed battery waste stream. Granata et al. [103] developed a hydro-pyrometallurgical process for a real upstream scrap containing mixed types of batteries. A fine fraction (−1000 μm) resulting from crushing and sieving spent LIBs was leached by a mixture of hydrochloric and nitric acid. Ni, Co, and Al were separated by solvent extraction. Nan et al. [104] also developed a process to recycle LiBs and NiMH batteries. A powder of mixed NiMH and LiB batteries was leached with H_2SO_4 and H_2O_2. REE elements were precipitated and separated from the solution. A solvent extractor (Acorga M5640) was used to extract the copper from the filtrate and the cobalt and nickel were extracted by Cyanex

272 at different pH. Dorella and Mansur [105] investigated a combined hydrometallurgical route consisting of manual dismantling, anode/cathode manual separation, acid leaching, chemical precipitation, and solvent extraction. Around 55 % of aluminum, 80 % of cobalt and 95 % of lithium were leached. Aluminum was partially separated from cobalt and lithium by precipitation at pH 5. After filtration, the aqueous solution was submitted to a purification step by liquid–liquid extraction with Cyanex 272 and around 85 % of cobalt was separated.

Cardarelli and Dubi [106] developed a novel method to recycle Li-polymer batteries (LMPBs). Vanadium is separated from the mother liquor as hydrated vanadium oxide by adjusting pH followed by precipitation and filtration. After removing aluminum by precipitation, purifying from heavy metals, and concentrating by evaporation, the final solution is left with purified lithium sulfate.

Contestabile et al. [93] used recycled cobalt hydroxide to synthesize lithium cobalt oxide and Kim et al. [107] reported that $LiCoO_2$ cathode material was renovated and simultaneously separated from spent electrodes containing $LiCoO_2$, electron-conducting carbon, binders, Al current collectors, and separator in a single synthetic step. The good working capabilities of the electrode were demonstrated and showed that the renovated $LiCoO_2$ phase is a promising cathode material for LIBs. Ra and Han [108] used Etoile–Rebatt technology to recover and renovate $LiCoO_2$ from LiBs scrap. The process includes the dissolution of spent electrode materials after which an electric current was used to precipitate the dissolved $LiCoO_2$ on a platinum electrode in a single synthetic step. The recovered and renovated $LiCoO_2$ exhibits a promising electrochemical activity and battery performance.

Attention should be paid to safety precautions in the recycling of LIBs. Unlike other batteries, LIBs often blow up during the recycling process due to radical oxidation of the lithium metal produced from battery metals. Metallic lithium can be deposited on the LiB current collectors due to overcharging/discharging and abnormal deposition [93]. Vigorous oxidation of metallic lithium in moisture or air can be dangerous during the mechanical treatment of spent LIBs, as $LiPF_6$ may react with water and produce toxic gases [101].

There are also some industrial scale recycling processes. One of the early processes developed for recycling LiBs is Toxco's process, where lithium carbonate (Li_2CO_3) is obtained as a final product [109, 110] The Recupyl process uses hydrolysis to obtain Li_2CO_3 from mixed spent LIBs [111].

In the Sumitomo-Sony technique, the batteries are calcined to burn out electrolytes and plastics and the remaining scrap is pyrometallurgically treated to recover Co–Ni–Fe. Cobalt is then hydrometallurgically separated from the alloy in order to reuse it in the manufacture of new batteries; copper and stainless steel are physically separated as by-product, while lithium and other valuable metals are lost in the slag [106].

The Inmteco process was not originally developed for battery recycling, but spent LiB and NiMH batteries are used as a secondary source with iron-containing materials in order to produce iron-based alloys. Copper and nickel are recovered while other metals are slagged. Carbon and organic materials are burnt and utilized as reducing agent [112]. The process from ACCUREC recovers a Li_2O concentrate by vacuum pyrolysis.

The Xstrata method from Canada combines a conditioning of batteries in a rotary kiln, introducing the product into a Co–/Ni winning process in an electric arc furnace. There is a subsequent hydrometallurgical step to recover Ni and Co in metallic form. This process does not recover Li [1]. In 2011, Umicore (Hoboken, Belgium) and Rhodia (La Rochelle, France) jointly developed a pyro-/hydrometallurgical process for the recycling of nickel metal hydride and Li-ion batteries [1].

10.4 Zinc-Manganese Dioxide Systems

Denise Crocce Romano Espinosa, Jorge Alberto Soares Tenório

Polytechnic School, USP, São Paulo, Brazil
e-mail: espinosa@usp.br

Alkaline and Zn-C (Leclanché cell) batteries, along with Zinc-Chloride batteries, are primary batteries and usually used in devices that require low voltages between 1.7 and 0.8 V. The Zn-C and the Zinc-Chloride batteries are also called Dry-batteries [113].

The battery converts stored chemical energy into electrical energy in the Zinc-Manganese Dioxide Systems according to the reaction of Eq. 10.8 [114, 115]:

$$Zn + 2MnO_2 \rightarrow ZnO + Mn_2O_3 \tag{10.8}$$

Zinc-C batteries are still one of the cheapest batteries due their simple assembly compared to others. The composition and configuration of these batteries are described in Fig. 10.3a: anode composed by an alloy of Zinc, with Lead (Pb) for improved malleability; Cadmium (Cd) and Mercury (Hg) as corrosion inhibitors; a cathode made of Manganese Dioxide, obtained from high purity ores, mixed with Carbon to increase conductivity and hold moisture; and with ammonium chloride and zinc chloride as electrolytes forming a thick paste in contact with a carbon conductor [113, 114].

The energy conversion is represented by Eq. 10.9, the characteristic Zn-C equation [116]:

$$2MnO_2 + 2NH_4Cl + Zn \rightarrow ZnCl_2 + 2NH_3 + Mn_2O_3 + H_2O \tag{10.9}$$

The Zinc-Chloride battery only differs from the Zn-C regarding its electrolyte, which is composed only by Zinc-Chloride, and because it has a lower content of Cd and Hg on the Zinc Alloy.

Table 10.2 [15] shows the average composition of dry batteries for the main manufactures.

Figure 10.3b shows the Alkaline battery scheme. The Alkaline battery follows the same general equation for its discharge, but its functioning and its composition are slightly different: an alkaline electrolyte, usually potassium hydroxide,

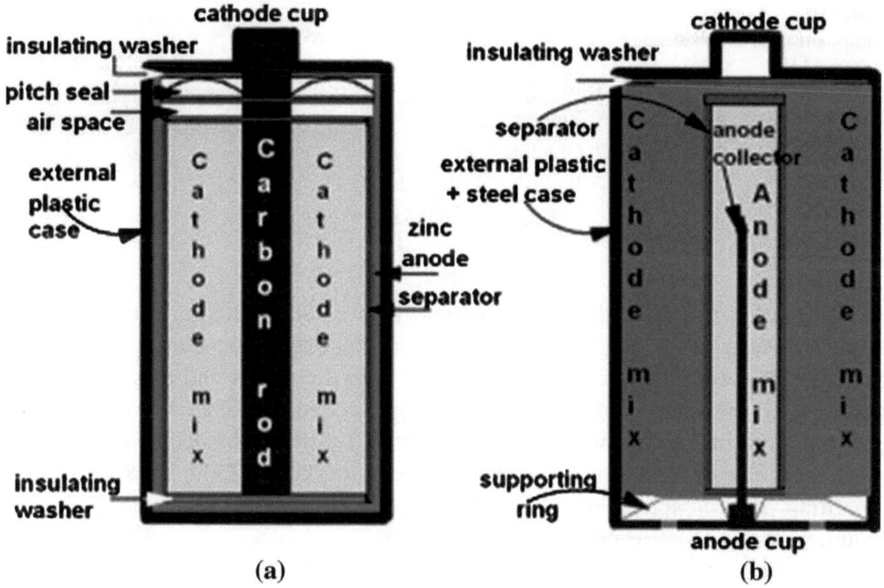

Fig. 10.3 Simplified Zinc-C battery (**a**) and an alkaline battery (**b**) [117]

Table 10.2 Average Mass composition of the dry battery (Zinc-chloride) [15]

Compound	Mass%
Steel	8–14
Manganese dioxide	28–32
Zinc	16–20
Graphite (carbon)	7–13
Zinc chloride	6–10
Water, paper and plastics	Balance

which ensures the corrosion protection for the Zinc, and a cathode composed of a pure manganese dioxide. They are also considered more environmentally friendly because of their lower rates of Hg, Cd and Pb. in addition to their a longer lifetime compared to similar batteries [118].

The assembly of these batteries using Zinc powder, ensures a higher surface area between anode and cathode, allowing a higher bulk density among the Zinc-Manganese dioxide systems. The specific alkaline equation (Eq. 10.10) ensures this property:

$$Zn + 2MnO_2 + H_2O + 2OH^- \rightarrow [Zn(OH)_4]^{2-} + Mn_2O_3 \qquad (10.10)$$

The average composition of alkaline batteries from the main manufacturers can be described as presented in Table 10.3 [119].

Table 10.3 Average composition of an alkaline battery [119]

Compound	Mass%
Steel	19–23
Manganese dioxide	32–38
Zinc	11–16
Graphite (carbon)	3–5
Potassium hydroxide	5–9
Water, paper and plastics	Balance

The market share for these alkaline and zinc carbon batteries currently represents 15 and 6 % of the global market [4], respectively, considering all kinds of batteries. However, they are usually cheaper than the others and they are primary batteries, so the same market share regarding mass and volume is quite the opposite, resulting in a higher waste production which justifies the necessity of an efficient recycling process. Based on the European market in 2009, the market share regarding mass is approximately 74 % (alkaline) and 20 % (zinc) and regarding volume, 70 and 18 % respectively [12].

According to a 2010 study of the European Batteries Association (EPBA), alkaline batteries have been increasing their market share since 1997 in comparison to Zinc-C and Zinc-Cl batteries [5]. So the majority of batteries found on household wastes is formed mostly by alkaline batteries.

Eventhough Zinc, Manganese and Iron (metals with higher concentrations in these batteries) have lower toxicities [11], the European Batteries Directive 2006/66/EC determines that a selective collection system is needed to treat the waste from both alkaline and zinc-carbon batteries [16]. Although some restrictions exist due the presence of mercury and lead in the batteries, the selective collection enables recycling them together due to their similar composition [120].

The initial phase of any recycling process starts with mineral processing techniques, most commonly physical stages that will concentrate metallic fractions and separate other materials such as plastics and scrap metals [19]. This tactic does not change the chemical composition of the material, it just physically prepares it for the next stage.

The most commonly used sequence used in such mineral processing techniques is:

- Sorting: this stage separates batteries with similar composition, avoiding contamination by other metals, such as mercury or cadmium. It is usually done through the selective gathering by public or private initiatives, or by manual sorting at a recycling center [121];
- Dismantling: this process is usually performed using knife or ball mills, which separate iron scraps and plastic particles from the black powder by cutting and crunching them. This technique uses coupled grills, magnets and fans to selectively gather particles with different physical properties, such as magnetic and conductivity properties, specific mass etc. [120, 121];
- Grinding: the purpose of this stage is to reduce particle size and create a homogeneous black powder, usually using a ball mill, increasing efficiency for further processing [121, 122].

After physical processing there are two routes used to recycle these batteries: a Hydrometallurgical route, based on aqueous acid/basic solutions and temperatures below 100 °C and; a Pyrometallurgical route, based on the use of high temperature furnaces, calcination and distillation of materials. Both process are capable of recovering the target metals in a highly pure form and they can be used together [121].

Figure 10.4 summarizes some of the basic sequences that are used to recover some metals present in recycled batteries.

Further details will be placed on each route, those are the most common processes industrially used.

Pyrometallurgical recycling includes the following basic techniques:

- *Pyrolysis*: this process evaporates water and mercury due its low boiling point (approx. 356 °C) [123], avoiding further contamination of the final products and recovering water. Also, residual organic materials are thermally decomposed into CO_2 and other gases [19];
- *Reduction*: the carbon generated in the previous process reduces the metallic compounds into alloys at temperatures of around 1,500 °C [19];
- *Incineration*: the resulting gas from the reduction and pyrolysis is incinerated and then quenched to avoid the generation of dioxins [19];
- *Distillation*: Because Zn and Hg have low boiling points, they evaporate and can be recovered by cooling the vapors formed during the previous processes [19].

Because the pyrometallurgical process uses high temperatures to separate, react and separate the target metals into final products or base alloys, which are used as primary alloys in forging, and because gases and vapors are generated, several environmentalists accuse it of being a large polluter contributing to the greenhouse effect.

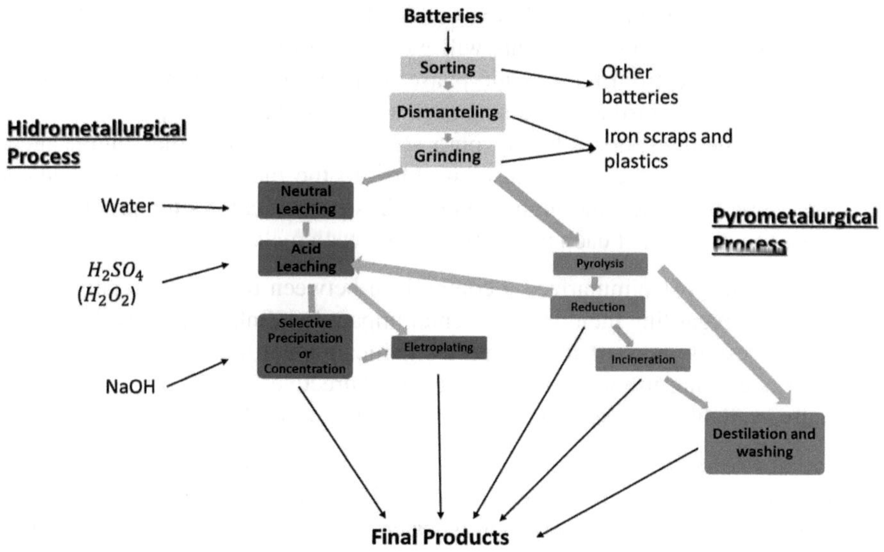

Fig. 10.4 Possible hydrometallurgical and/or pyrometallurgical processes [16, 19, 121]

The alternative process is the **hydrometallurgical recycling** of batteries, which includes complex methods thar are usually performed at temperatures below 100 °C and without the generation of fumes or gases. A common sequence used to recycle these batteries is:

- *Neutral Leaching*: this stage removes any soluble substance, such as electrolytes (Na and K), by using water. The recycling of alkaline batteries requires this if an electroplating stage will be performed because the high concentration of electrolytes can disturb electrowinning stages. The process does not affect the extraction of any target metals (Fe, Zn and Mn) because they are not soluble in basic solutions [121];
- *Acid Leaching*: the solid residue from the previous process is leached again with an acid solution of sulfuric acid (pH inferior to 1.0) to extract the target metals: Manganese, Zinc and Iron [124]. Zinc and Iron are soluble at low pH values, but if a soluble form of manganese is desirable in the final solution, H_2O_2 is crucial to reduce the manganese to a soluble form. Note that the reductive capacity of H_2O_2 in this case may influence the efficiency of zinc and iron leaching [121];
- *Concentration*: in this stage, the liquor produced is used to concentrate the target metals in different solutions, separating or precipitating them selectively. This may be done by a series of techniques, such as:
  - Selective precipitation: by increasing the pH of the liquor obtained from the acid leaching, metals will precipitate in a specific order according to their solid-liquid equilibria (found in the Pourbaix Diagrams). In this case, the sequence of precipitation is Fe, Zn an finally Mn [35].
  - Solvent extraction: this process uses extractants/molecules soluble in organic phases (usually kerosene) to selectively recover each metal, in its soluble form, from the complex solution obtained from the acid leaching. With this technique, different extractants will gather a specific metal, that will be further recovered from the organic phase on a different solution. This solution will be used on further processing such as electroplating [125].
  - Electrowinning: after a highly pure solution is obtained, the soluble metal is recovered by the reduction of its Me^+ to the metal state on the cathode by using electrodes and high voltages. This process allows the production of pure metal plates of each target metal from each metal solution.

Briffaerts et al. [16] summarized a comparison between the commercially available processes regarding their environmental impact for Zink-C and alkaline batteries and the results shows that none has a significant environmental advantage over the other according to the European Batteries Directive 2006/66/EC. Some commercially available processes to recycle Zinc-Manganese Dioxide batteries are:

- Revabat (Revatech in Belgium): this is mainly a mechanical process and most of the manganese and zinc ends up in the black powder, removing iron scraps and plastics that are sold separately to iron dealers and as refuse derived fuel in cement kilns. The polluted paper obtained is landfilled. The black powder containing manganese and zinc is further processed in chemical hydrometallurgical stages [16];

- Revatech (Belgium): hydrometallurgical process to valorize the black mass produced by Revabat. It produces manganese dioxide (98 % purity in dry matter) and zinc sulphate, and until 2009 the process was still on laboratory scale [16];
- Valdi (France): recycles batteries with a maximum 500 ppm concentration of Mercury, the treatment involves a pyrometallurgical step in electric arc furnaces (temperatures above 1,500 °C) to produce ferromanganese (containing nickel) and filter dust (zinc oxide) as final products without the generation of a large amount of waste products [16];
- Batrec (part of Veolia Group): developed to safely treat battery waste regardless of the mercury content. The process uses a pyrometallurgical stage in a shaft furnace at approx. 600 °C with several cleaning systems to safely recover and purify water and gases. After this stage, the residue is melted in an electrically heated induction melting furnace producing and uniform 40 % Mn ferromanganese alloy and slag [16].

Another process is the Recupyl method, which applies hydrometallurgical techniques for metals recovery [1].

References

1. Worrell E, Reuter MA (eds) (2014). Handbook of recycling. Elsevier, Amsterdam
2. Putois F (1995) Market for nickel-cadmium batteries. J Power Sources 57(1–2):67–70. doi:10.1016/0378-7753(95)02243-0
3. Buchman I (2001) Batteries in a portable world: a handbook on rechargeable batteries for non-engineers. 2nd edn. Cadex Eletronics Inc: 148
4. Buchmann I (2013) Global battery markets information—battery University. Avaliable on http://batteryuniversity.com/learn/article/global_battery_markets. Accessed 25 Apr 2013
5. Epba (2010) European Portable Battery Association. Epba sustainability Report
6. Vassura I (2009) Chemical characterisation of spent rechargeable batteries. Waste Manag 29(8):2332–2335. doi:10.1016/j.wasman.2009.03.033
7. Nogueira CAE, Margarido F (2007) Chemical and physical characterization of electrode materials of spent sealed Ni-Cd batteries. Waste Manag 27(11):1570–1579. doi:10.1016/j.wasman.2006.10.007
8. Radio Shack Corporation. Nickel-Cadmium batteries. Available on http://support.radioshack.com/support_tutorials/batteries/bt-nicd-main.htm. Accessed 10 Dec 2013
9. Schumm BJ (2012) Battery Electronics—Britannica Online Encyclopedia. Available on http://www.britannica.com/EBchecked/topic/56126/battery/45858/Development-of-batteries. Accessed 10 Dec 2013
10. Espinosa DCR, Bernardes AM, Tenório JAS (2004) Brazilian policy on battery disposal and its practical effects on battery recycling. J Power Sources 137(1):134–139. doi:10.1016/j.jpowsour.2004.02.023
11. Habeck M (2013) Eco-USA toxic chemicals. Available on http://www.eco-usa.net/toxics/index.shtml. Accessed 12 Mar 2013
12. Buchmann I (2013) Battery University-battery developments. Available on http://batteryuniversity.com/learn/article/battery_developments. Accessed 2 Mar 2013
13. Espinosa DCR, Bernardes AM, Tenório JAS (2004) An overview on the current processes for the recycling of batteries. J Power Sources 135:311–319. doi:10.1016/j.jpowsour.2004.03.083

14. Ruffino B, Zanetti MC, Marini P (2011) A mechanical pre-treatment process for the valorization of useful fractions from spent batteries. Resour Conserv Recycl 55(3):309–315. doi:10.1016/j.resconrec.2010.10.002
15. Pereira RSD (2012) Characterization of spent batteries collected in a recycle bin. Revista Brasileira de Ciências Ambientais 25(1):60–67. doi:10.1016/j.jpowsour.2004.02.023
16. Briffaerts K (2009) Waste battery treatment options: comparing their environmental performance. Waste Manag 29(8):2321–2331. doi:10.1016/j.wasman.2009.03.019
17. Huber ML, Laesecke A, Friend DG (2006) The vapor pressure of mercury. Available on http://www.physics.rutgers.edu/~eandrei/389/NISTIR.6643.pdf. Accessed 13 Jun 2014
18. Bernardes AM, Espinosa DCR, Tenório JAS (2004) Recycling of batteries: a review of current processes and technologies. J Power Sources 130(1–2):291–298. doi:10.1016/j.jpowsour.2003.12.026
19. Espinosa DCR, Tenório JAS (2004) Fundamental aspects of recycling of nickel–cadmium batteries through vacuum distillation. J Power Sources 135(1–2):320–326. doi:10.1016/j.jpowsour.2004.03.082
20. Velgosová O (2013) Influence of H2SO4 and ferric iron on Cd bioleaching from spent Ni–Cd batteries. Waste Manag 33(2):456–461. doi:10.1016/j.wasman.2012.10.007
21. Fernandes A, Afonso JC, Dutra AJB (2012) Hydrometallurgical route to recover nickel, cobalt and cadmium from spent Ni–Cd batteries. J Power Sources 220:286–291. doi:10.1016/j.jpowsour.2012.08.011
22. Ruetschi P, Meli F, Desilvestro J (1995) Nickel-metal hydride batteries. The preferred batteries of the future? J Power Sources 57:85–91
23. Lyman PW, Palmer GR (1993) Investigating the recycling of nickel hydride battery scrap. J Metals May. 32–35
24. Zhang P, Yokoyama T, Itabashi O, Wakui Y, Suzuki TM, Inoue K (1998) Hydrometallurgical process for recovery of metals values from spent nickel-metal hydride secondary batteries. Hydrometallurgy 50:61–75
25. Jossen A, Garche J, Sauer DU (2004) Operations conditions of batteries in Pv applications. Sol Energy 76:759–769
26. Hadjean RB, Ramos JPP, Latroche M, Guégan AP (2003) In Sity X-ray diffraction study on MmNi3.85Mn0.27Al0.37Co0.38 as negative electrode in alkaline secondary batteries. Electrochim Acta 48:2813–2821
27. Gschneidner KA, Beaudry BJ, Capellen J (1990) Properties and selection: nonferrous alloys and special-purpose materials. Vol. 2, ASM Handbook. ASM International ISBN 978-0-87170-378-1
28. Baumann W, Muth A (1997) Batterien – Daten Und Fakten Zum Umweltschutz. Springer. ISBN 3-540-61594-6
29. Müller T, Friedrich B (2006) Development of a recycling process for nickel-metal hydride batteries. J Power Sources 158:1498–1509
30. Nan J, Han D, Yang M, Cui M, Hou X (2006) Recovery of metal values from a mixture of spent lithium-ion batteries and nickel-metal hydride batteries. Hydrometallurgy 84:75–80
31. Bertuol DA, Bernardes AM, Tenório JAS (2006) Spent NiMH batteries: characterization and metal recovery through mechanical processing. J Power Sources 160:1465–1470
32. Huang K, Li J, Xu Z (2011) Enhancement of the recycling of waste Ni–Cd and Ni–MH batteries by mechanical treatment. Waste Manag 31:1292–1299
33. Al-Thyabat S, Nakamura T, Shibata E, Iizuka A (2013) Adaptation of minerals processing operations for lithium-ion (LiBs) and nickel metal hydride (NiMH) batteries recycling: critical review. Miner Eng 45:4–17
34. Binnemans K, Jones PT, Blanpain B, Gerven TV, Yang Y, Walton A, Buchert M (2013) Recycling of rare earths: a critical review. J Clean Prod 51:1–22
35. Bertuol DA, Bernardes AM, Tenório JAS (2009) Spent NiMH batteries batteries—the role of selective precipitation in the recovery of valuable metals. J Power Sources 193:914–923
36. Hoogerstraete TV, Binnemans K (2014) Highly efficient separation of rare earths from nickel and cobalt by solvent extraction with the ionic liquid trihexyl(tetradecyl)phosphonium nitrate: a process relevant to the recycling of rare earths from permanent magnets and nickel metal hydride batteries. Green Chem 16:1594–1606

37. Rodrigues LEOC, Mansur MB (2010) Hydrometallurgical separation of rare earth elements, cobalt and nickel from spent nickel–metal–hydride batteries. J Power Sources 195:3735–3741
38. Li L, Xu S, Ju Z, Wu F (2009) Recovery of Ni, Co and rare earths from spent Ni–metal hydride batteries and preparation of spherical $Ni(OH)_2$. Hydrometallurgy 100:41–46
39. Innocenzi V, Veglio F (2012) Separation of manganese, zinc and nickel from leaching solution of nickel-metal hydride spent batteries by solvent extraction. Hydrometallurgy 129–130:50–58
40. Larsson K, Ekberg C, Odegaard-Jensen A (2013) Using Cyanex 923 for selective extraction in a high concentration chloride medium on nickel metal hydride battery waste. Part II: mixer–settler experiments. Hydrometallurgy 133:168–175
41. Gasser MS, Aly MI (2013) Separation and recovery of rare earth elements from spent nickel–metal-hydride batteries using synthetic adsorbent. Int J Miner Process 121:31–38
42. Bertuol DA, Amado FDR, Veit H, Ferreira JZ, Bernardes AM (2012) Recovery of nickel and cobalt from spent NiMH batteries by electrowinning. Chem Eng Technol 35(12):2084–2092
43. Peng B, Chen J (2009) Functional materials with high-efficiency energy storage and conversion for batteries and fuel cells. Coord Chem Rev 253(23–24):2805–2813
44. Zhang T, Li D, Tao Z, Chen J (2013) Understanding electrode materials of rechargeable lithium batteries via DFT calculations. Prog Nat Sci Mater Int 23(3):256–272
45. Al-Thyabat S, Nakamura T, Shibata E, Iizuka A (2013) Adaptation of minerals processing operations for lithium-ion (LiBs) and nickel metal hydride (NiMH) batteries recycling: Critical review. Miner Eng 45:4–17
46. Nagaura T, Tozawa K (1990) Progress in batteries and solar cells. JEC Press 9:209
47. Nishi Y (2001) Lithium ion secondary batteries; past 10 years and the future. J Power Sources 100:101–106
48. Kucinskis G, Bajars G, Kleperis J (2013) Graphene in lithium ion battery cathode materials: a review. J Power Sources 240:66–79
49. Barré A, Deguilhem B, Grolleau S, Gérard M, Suard F, Riu D (2013) A review on lithium-ion battery ageing mechanisms and estimations for automotive applications. J Power Sources 241:680–689
50. Castillo S, Ansart F, Laberty-Robert C, Portal J (2002) Advances in the recovering of spent lithium battery compounds. J Power Sources 112:247–254
51. Wang RC, Lin YC, Wu SH (2009) A novel recovery process of metal values from the cathode active materials of the lithium-ion secondary batteries. Hydrometallurgy 99:194–201
52. Xu J, Thomas HR, Francis RW, Lum KR, Wang J, Liang B (2008) A review of processes and technologies for the recycling of lithium-ion secondary batteries. J Power Sources 177:512–527
53. Mukherjee R, Krishnan R, Lu TM, Koratkar N (2012) Nanostructured electrodes for high-power lithium ion batteries. Nano Energy 1:518–533
54. Huang R, Ikuhara Y (2012) STEM characterization for lithium-ion battery cathode materials. Curr Opin Solid State Mater Sci 16:31–38
55. Georgi-Maschlera T, Friedrich B, Weyheb R, Heegnc H, Rutz M (2012) Development of a recycling process for Li-ion batteries. J Power Sources 207:173–182
56. Whittingham MS (2004) Lithium batteries and cathode materials. Chem Rev 104:4271–4301
57. Purkayastha R, McMeeking R (2013) A parameter study of intercalation of lithium into storage particles in a lithium-ion battery. Comput Mat Sci. http://dx.doi.org/10.1016/j.commatsci.2012.11.050
58. Zhang P, Yokoyama T, Itabashi O, Suzuki TM, Inoue K (1998) Hydrometallurgical process for recovery of metal values from spent lithium-ion secondary batteries. Hydrometallurgy 47:259–271
59. Scrosati B, Garche J (2010) Lithium batteries: status, prospects and future. J Power Sources 195:2419–2430

60. Fergus JW (2010) Recent developments in cathode materials for lithium ion batteries. J Power Sources 195:939–954
61. Mohan VM, Hu B, Qiu W, Chen W (2009) Synthesis, structural, and electrochemical performance of V_2O_5 nanotubes as cathode material for lithium battery. J Appl Electrochem 39:2001–2006
62. Hao Q, Ma H, Ju Z, Li G, Li X, Xu L, Qian Y (2011) Nano-CuO coated $LiCoO_2$: synthesis, improved cycling stability and good performance at high rates. Electro-chimica Acta 56(25):9027–9031
63. Xia H, Luo Z, Xie J (2012) Nanostructured $LiMn_2O_4$ and their composites as high-performance cathodes for lithium-ion batteries. Prog Nat Sci Mat Int 22:572–584
64. Zhang Y, Huoc QY, Dua P, Wanga L, Zhang A, Song Y, Lv Y, Li G (2012) Advances in new cathode material $LiFePO_4$ for lithium-ion batteries. Synth Met 162:1315–1326
65. Casas C, Li W (2012) A review of application of carbon nanotubes for lithium ion battery anode material. J Power Sources 208:74–85
66. Wu H, Cui Y (2012) Designing nanostructured Si anodes for high energy lithium ion batteries. Nano Today 7:414–429
67. Howe JY, Burton DJ, Qi Y, Meyer HM, Nazri M, Nazri GA, Palmer AC, Lake PD (2013) Improving microstructure of silicon/carbon nanofiber composites as a Li battery anode. J Power Sources 221:455–461
68. Zhanga M, Gao B, Li Y, Zhang X, Hardin IR (2013) Graphene-coated pyrogenic carbon as an anode material for lithium battery. Chem Eng J 229:399–403
69. Tang Q, Shan Z, Wangb L, Qin X (2012) MoO_2–graphene nanocomposite as anode material for lithium-ion batteries. Electrochim Acta 79:148–153
70. Kim KM, Kang KY, Kim S, Lee YG (2012) Electrochemical properties of TiO_2 nanotube-$Li_4Ti_5O_{12}$ composite anodes for lithium-ion batteries. Curr Appl Phys 12:1199–1206
71. Wang J, Zhao H, Yang Q, Wang C, Lv P, Xia Q (2013) $Li_4Ti_5O_{12}$ -TiO_2 composite anode material for lithium-ion batteries. J Power Sources 222:196–201
72. Kim Y, Yoon Y, Shin D (2009) Fabrication of Sn/SnO_2 composite powder for anode of lithium ion battery by aerosol flame deposition. J Anal Appl Pyrolysis 85:557–560
73. Rai AK, Anh LT, Gim J, Mathew V, Kim J (2013) Low temperature synthesis of porous tin oxide anode for high-performance lithium-ion battery. Electrochimica Acta. http://dx.doi.org/10.1016/j.electacta.2013.07.128
74. Schedlbauer T, Krüger S, Schmitz R, Schmitz RW, Schreiner C, Gores HJ, Passerini S, Winter M (2013) Lithium difluoro(oxalato)borate: a promising salt for lithium metal based secondary batteries? Electrochim Acta 92:102–107
75. Di Leo RA, Marschilok AC, Takeuchi KJ, Takeuchi ES (2013) Battery electrolytes based on saturated ring ionic liquids: physical and electrochemical properties. Electrochim Acta 109:27–32
76. Li J, Daniel C, Wood D (2011) Materials processing for lithium-ion batteries. J Power Sources 196:2452–2460
77. Tatsumisago M, Nagao M, Hayashi A (2013) Recent development of sulfide solid electrolytes and interfacial modification for all-solid-state rechargeable lithium batteries. J Asian Ceram Soc 1:17–25
78. Takada K (2013) Progress and prospective of solid-state lithium batteries. Acta Mater 61:759–770
79. Idris NH, Rahman M, Wang JZ, Liu HK (2012) Microporous gel polymer electrolytes for lithium rechargeable battery application. J Power Sources 201:294–300
80. Rupp B, Schmuck M, Balducci A, Winter M, Kern W (2008) Polymer electrolyte for lithium batteries based on photochemically crosslinked poly(ethylene oxide) and ionic liquid. Eur Polymer J 44:2986–2990
81. Prasanth R, Shubha N, Hoon HH, Srinivasan M (2013) Effect of nano-clay on ionic conductivity and electrochemical properties of poly(vinylidene fluoride) based nanocomposite porous polymer membranes and their application as polymer electrolyte in lithium ion batteries. Eur Polymer J 49:307–318

82. Rao M, Geng X, Liao Y, Hu S, Li W (2012) Preparation and performance of gel polymer electrolyte based on electrospun polymer membrane and ionic liquid for lithium ion battery. J Membr Sci 399–400:37–42
83. Hofmann A, Schulz M, Hanemann T (2013) Gel electrolytes based on ionic liquids for advanced lithium polymer batteries. Electrochim Acta 89:823–831
84. Huang X (2012) A lithium-ion battery separator prepared using a phase inversion process. J Power Sources 216:216–221
85. Costa CM, Ribelles JLG, Lanceros-Méndez S, Appetecchi GB, Scrosati B (2014) Poly(vinylidene fluoride)-based, co-polymer separator electrolyte membranes for lithium-ion battery systems. J Power Sources 245:779–786
86. Huang X, Hitt J (2013) Lithium ion battery separators: development and performance characterization of a composite membrane. J Membr Sci 425–426:163–168
87. Jeong HS, Choi ES, Lee SY, Kim JH (2012) Evaporation-induced, close-packed silica nanoparticle-embedded nonwoven composite separator membranes for high-voltage/high-rate lithium-ion batteries: advantageous effect of highly percolated, electrolyte-philic microporous architecture. J Membr Sci 415–416:513–519
88. Mantuano DP, Dorella G, Elias RCA, Mansur MB (2006) Analysis of a hydrometallurgical route to recover base metals from spent rechargeable batteries by liquid–liquid extraction with Cyanex 272. J Power Sources 159. http://dx.doi.org/10.1016/j.jpowsour.2005.12.056
89. Castillo S, Ansart F, Laberty CR, Portal J (2002) Advances in the recovering of spent lithium battery compounds. J Power Sources 112:247–254
90. Saeki S, Lee J, Zhang QW, Saito F (2004) Co-grinding LiCoO2 with PVC and water leaching of metal chlorides formed in ground product. Int J Miner Process 74S:373–378
91. Xu J, Thomas HR, Francis RW, Lum KR, Wang J, Liang B (2008) A review of processes and technologies for the recycling of lithium-ion secondary batteries. J Power Sources 177:512–527
92. Lee CK, Rhee KI (2003) Reductive leaching of cathodic active materials from lithium ion battery wastes. Hydrometallurgy 68:5–10
93. Contestabile M, Panero S, Scrosati B (2001) A laboratory-scale lithium-ion battery recycling process. J Power Sources 92:65–69
94. Kawakami S (1999) Method for recovering lithium cell materials. US Patent 5882811
95. Lin J, Fan C, Chang I, Shiu J (2003) Clean process of recovering metals from waste lithium ion batteries. USPatent 6514311B1
96. Mishra D, Kim D, Ralph D, Ahn J, Rhee Y (2008) Bioleaching of metals from spent lithium ion secondary batteries using Acidithiobacillus ferrooxidans. Waste Manag 28:333–338
97. Chen M, Zhang F, Zhu J (2011) A preliminary investigation for spent LiBs recycling. In: Supplemental Proceedings Materials Processing and Energy Materials vol.1. TMS (The minerals, metals & materials society): 25–32
98. Castillo S, Ansart F, Laberty-Robert C, Portal J (2002) Advances inthe recovering of spent lithium battery compounds. J Power Sources 112:247–254
99. Myoung J, Jung Y, Lee J, Tak Y (2002) Cobalt oxide preparation from waste LiCoO2 by electrochemical–hydrothermal method. J Power Sources 112: 639–642
100. Shen Y (2002) Chinese Nonferr. Metals 54:69–71 (in Chinese)
101. Lupi C, Pasquali M (2003) Electrolytic nickel recovery from lithium-ion batteries. Miner Eng 16:537–542
102. Lain MJ (2001) Recycling of lithium ion cells and batteries. J Power Sources 97(98):736–738
103. Granata G, Pagnanelli F, Moscardini E, Takacova Z, Havlik T, Toro L (2012) Simultaneous recycling of nickel metal hydride, lithium ion and primary lithium batteries: accomplishment of European guidelines byoptimizing mechanical pre-treatment and solvent extraction operations. J Power Sources 212:205–211
104. Nan J, Han D, Yang M, Cui M, Hou X (2006) Recovery of metal values from a mixture of spent lithium-ion batteries and nickel–metal hydride batteries. Hydrometallurgy 84:75–80
105. Dorella G, Mansur MB (2007) A study of the separation of cobalt from spent Li-ion battery residues. J Power Sources 170:210–215

106. Cardarelli F, Dubi J (2007) Method for recycling spent lithium metal polymer rechargeable batteries and related materials. USPatent No.7192564B2
107. Kim DS, Sohn JS, Lee CK, Lee JH, Hanc KS, Lee YI (2004) Simultaneous separation and renovation of lithium cobalt oxide from the cathode of spent lithium ion rechargeable batteries. J Power Sources 132:145–149
108. Ra D, Han K (2006) Used lithium ion rechargeable battery recycling using Etoile–Rebatt technology. J Power Sources 163:284–288
109. Maclaughlin W (1994) Method for neutralization of hazardous materials. US Patent 5345033
110. Maclaughlin W, Adams T (1999) Lithium reclamation process. US Patent 5888463
111. Tedjar F, Foudraz J (2010) Method for the mixed recycling of lithium-based anode batteries and cells. US Patent 7820317B2
112. Georgi-Maschlera T, Friedricha B, Weyheb R, Heegnc H, Rutzc M (2012) Adaptation of minerals processing operations for lithium-ion (LiBs) and nickel metal hydride (NiMH) batteries recycling: critical review. J Power Sources 207:173–182
113. Britannica Online Encyclopedia. Battery (electronics). Available on www.britannica.com/EBchecked/topic/56126/battery?sections=56126TO. Accessed 3 Nov 2011
114. Shin SM (2009) Separation of zinc from spent zinc-carbon batteries by selective leaching with sodium hydroxide. Hydrometallurgy 96(4):349–353. doi:10.1016/j.hydromet.2008.12.010
115. Doitpoms—TLP Library—Examination of a Manufactured Article—Polymers. Available on http://www.doitpoms.ac.uk/tlplib/artefact/polymers.php
116. Eveready Battery Co. Inc. (2001) Eveready carbon Zinc (Zn/MnO^2) application manual
117. Belardi G, Ballirano P, Ferrini M, Lavecchia R, Medici F, Piga L, Scoppettuolo A (2011) Characterization of spent zinc–carbon and alkaline batteries by SEM-EDS, TGA/DTA and XRPD analysis. Thermochim Acta 526(1–2):169–177
118. Schumm JB (2013) Battery Electronics—Britannica Online Encyclopedia. Available on http://www.britannica.com/EBchecked/topic/56126/battery/45858/Development-of-batteries. Accessed 10 Dec 2013
119. Provazi K, Campos BA, Espinosa DC, Tenório JA (2011) Metal separation from mixed types of batteries using selective precipitation and liquid-liquid extraction techniques. Waste Manag 31(1):59–64. doi:10.1016/j.wasman.2010.08.021
120. Souza CCB, Oliveira DC, Tenório JAS (2001) Characterization of used alkaline batteries powder and analysis of zinc recovery by acid leaching. J Power Sources 103:120–126
121. Veloso LRS (2005) Development of a hydrometallurgical route for the recovery of zinc and manganese from spent alkaline batteries. J Power Sources 152:295–302. doi:10.1016/j.jpowsour.2005.03.133
122. Espinosa DCR (2008) Reciclagem de Pilhas e Baterias. Universidade de São Paulo
123. Huber ML, Friend DG (2006) The vapor pressure of mercury
124. Bertin FCH (2014) Optimal leaching on hydrometallurgical process of recycling. In: TMS Annual Meeting and Exhibition, San Diego, CA
125. Saçmaci S, Kartal S (2008) Selective extraction, separation and speciation of iron in different samples using 4-acetyl-5-methyl-1-phenyl-1H-pyrazole-3-carboxylic acid. Anal Chim Acta 623(1):46–52. doi:10.1016/j.aca.2008.05.073

MIX
Papier aus verantwortungsvollen Quellen
Paper from responsible sources
FSC® C105338

If you have any concerns about our products,
you can contact us on
ProductSafety@springernature.com

In case Publisher is established outside the EU,
the EU authorized representative is:
Springer Nature Customer Service Center GmbH
Europaplatz 3, 69115 Heidelberg, Germany

Printed by Libri Plureos GmbH
in Hamburg, Germany